APR 1 1 2014

WHAT IS RELATIVITY?

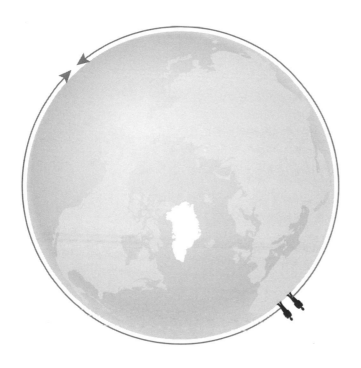

ALSO BY JEFFREY BENNETT

FOR ADULTS

*On the Cosmic Horizon: Ten Great Mysteries
    for Third Millennium Astronomy*

*Beyond UFOs: The Search for Extraterrestrial Life and
    Its Astonishing Implications for Our Future*

Math for Life: Crucial Ideas You Didn't Learn in School

FOR CHILDREN

Max Goes to the Moon

Max Goes to Mars

Max Goes to Jupiter

Max Goes to the Space Station

The Wizard Who Saved the World

TEXTBOOKS

*Using and Understanding Mathematics:
    A Quantitative Reasoning Approach*

Statistical Reasoning for Everyday Life

Life in the Universe

The Cosmic Perspective

The Essential Cosmic Perspective

The Cosmic Perspective Fundamentals

JEFFREY BENNETT

WHAT IS RELATIVITY?

AN INTUITIVE INTRODUCTION TO EINSTEIN'S IDEAS, AND WHY THEY MATTER

COLUMBIA UNIVERSITY PRESS
NEW YORK

Columbia University Press
Publishers Since 1893
New York   Chichester, West Sussex
cup.columbia.edu
Copyright © 2014 Jeffrey Bennett

Figures 1.1, 1.3, 2.1–2.4, 3.1–3.5, 4.1, 4.2, 5.1–5.6, 6.2–6.5, 6.8–6.11, 7.1, and 8.1–8.3 are all adapted adapted from similar illustrations in Jeffrey Bennett, Megan Donahue, Nick Schneider, and Mark Voit, *The Cosmic Perspective*, 7th ed. (2014). By permission of Pearson Education, Inc., Upper Saddle River, N.J.

Library of Congress Cataloging-in-Publication Data
Bennett, Jeffrey O.
   What is relativity? : an intuitive introduction to einstein's ideas, and why they matter / Jeffrey Bennett.
       pages cm
   Includes index.
   ISBN 978-0-231-16726-0 (cloth : alk. paper) —
   ISBN 978-0-231-53703-2 (e-book)
   1. Relativity (physics)—Popular works.   I. Title.
QC173.57.B46 2014
530.11—dc23

2013026801

Columbia University Press books are printed on permanent and durable acid-free paper.
This book is printed on paper with recycled content.
Printed in the United States of America

c 10 9 8 7 6 5 4 3 2 1

Cover image: © Bettmann/CORBIS. Personality rights of Albert Einstein are used with permission of the Hebrew University of Jerusalem.
Cover design: Alex Camlin
Book design: Lisa Hamm

References to websites (URLs) were accurate at the time of writing.
Neither the author nor Columbia University Press is responsible for URLs that may have expired or changed since the manuscript was prepared.

CONTENTS

In celebration of the 100th anniversary of the 1915 publication of Einstein's general theory of relativity.

PREFACE

MY FIRST real exposure to Einstein's theory of relativity came in a course I took during my freshman year of college. Like everyone else, I'd always heard that relativity was supposed to be really hard. But as I listened to my professor and studied at home, I soon realized that its reputation was undeserved. Relativity didn't make things harder; it made everything seem simpler, at least once you got the hang of it. It also seemed important—I suddenly realized that prior to studying relativity, I had misunderstood the basic nature of space and time. Given that we spend our entire lives living on a planet in space and moving through time, that seemed like a rather fundamental gap in my earlier education.

Within a year I was teaching some of the ideas of relativity to elementary and middle school children, as part of a summer school I ran for kids who were interested in space and science. I was amazed at how readily many of them grasped the key ideas, and their ease with the concepts helped me realize a fundamental fact: Much of the difficulty that most people have with relativity comes about only because it seems to run counter to ideas of space and time that have become deeply ingrained in our minds. For children, who have those ideas less deeply ingrained, relativity does not seem quite so strange, allowing them to accept it more easily than adults.

This insight proved particularly helpful a few years later, when I began teaching at the college level and made relativity an integral part of my courses in introductory astronomy. On the basis of my work with younger children, I focused on helping students overcome their natural resistance to revising their ideas of space and time. An added advantage of this approach was that it could be done without most of the mathematics that goes with relativity, allowing the students to focus on building a conceptual understanding. Year after year, on end-of-course evaluations, relativity consistently took the top spot when I asked students their favorite part of the course. When I asked why they enjoyed relativity so much, the most commonly cited reasons were (1) they appreciated the way relativity opened their minds in new and unexpected ways; and (2) they'd always assumed that relativity was a subject that would be beyond their comprehension, so they were excited to find out that they could actually understand it.

Over the years, I continued to emphasize relativity in my astronomy classes, and continued to refine my approach to teaching it. When three friends (Mark Voit, Megan Donahue, and Nick Schneider) and I landed a contract to write a textbook for introductory astronomy, we included two full chapters on relativity, even though surveys showed that very few faculty spent significant time teaching relativity in their astronomy courses for nonscience majors. We have at least some evidence that our inclusion of these chapters has inspired more instructors to include the topic.

That brings us to my goals for this book. I hope to help you, the reader, gain the same kind of appreciation for relativity that I have gained myself, and that I hope I've shared with my past students and with readers of my textbooks. I think you'll find the subject to be both much easier to understand and much more amazing than you expected. I also hope you'll come to agree with me that relativity is *important* to the way we view ourselves as human beings in a vast universe. As we approach the 100th anniversary of Einstein's 1915 publication of his general theory of relativity, I believe it's time to take relativity out of the realm of obscure science and bring it into the realm of general public consciousness. If this book helps accomplish that, then I will feel it has been a success.

Jeffrey Bennett

Boulder, Colorado

WHAT IS RELATIVITY?

Part 1

GETTING STARTED

1

VOYAGE TO A BLACK HOLE

IMAGINE THAT the Sun magically collapsed, retaining the same mass but shrinking in size so much that it became a black hole. What would happen to Earth and the other planets? Ask almost anyone, including elementary school kids, and they'll tell you confidently that the planets "would be sucked in."

Now imagine that you're a future interstellar traveler. Suddenly, you discover that a black hole lurks off to your left. What should you do? Again, ask around, and you'll probably be told to fire up your engines to try to get away, and that you'll be lucky to avoid being "sucked into oblivion."

But I'll let you in on a little secret that's actually important to understanding relativity: *Black holes don't suck*. If the Sun suddenly became a black hole, Earth would become very cold and dark. However, since we've assumed that the black hole will have the same mass as the Sun, Earth's orbit would hardly be affected at all.

As for your future as an interstellar traveler . . . First of all, you wouldn't "suddenly" discover a black hole off to your left. We have ways to detect many black holes even from Earth, and if we are someday able to embark on interstellar trips we'll surely have maps that would alert you to the locations of any black holes along your route. Even in the unlikely event that one wasn't on your map, the black hole's gravitational effect on your spacecraft would build gradually as you approached, so there'd still be nothing sudden about it. Second, unless you happened to be aimed almost directly at the black hole, its gravity would simply cause you to swing around it in much the same way that we've sent spacecraft (such as the *Voyager* and *New Horizons* spacecraft) swinging past Jupiter on trips to the outer solar system.

I realize that this may be very disappointing to some of you. As my middle-school daughter put it, "But it's cool to think that black holes suck." I was able to placate her only somewhat by pointing out that being cool and "it sucks" don't usually go together. Still, you're probably wondering, if black holes don't suck, what *do* they do?

The answer has two parts, one mundane and one so utterly amazing that you'll never again miss your visions of a cosmic vacuum cleaner. The mundane part applies to black holes observed from afar, because at a distance the gravity of a black hole is no different than the gravity of any other object. That's why turning the Sun into a black hole would not affect Earth's orbit, and why a spacecraft can swing by a black hole just like it swings by Jupiter. The amazing part comes when you begin to approach a black hole closely. There, you'd begin to observe the dramatic distortions of space and time that we can understand only through Einstein's *theory of relativity*.

That brings us to the crux of the matter. I've begun this book on relativity by talking about black holes because although almost everyone has heard of them, you cannot actually understand what black holes are unless you first understand the basic ideas discovered by Einstein. One goal of this book is to help you gain that understanding. But I have a second, more important goal in mind as well.

In the process of learning about relativity, you'll find that your everyday notions of time and space do not accurately reflect the reality of the universe. In essence, you'll realize that you have grown up with a "common sense" that isn't quite as sensible as it seems. It's not your fault; rather, it is a result of the fact that we don't commonly experience the extreme conditions under which the true nature of time and space is most clearly revealed. Therefore, the real goal of this book is to help you to distinguish reality from the fiction that we grow up with, and in the process to consider some of the profound implications of this reality that Einstein was the first to understand.

To get started, let's take an imaginary voyage to a black hole. This journey will give you an opportunity to experience the two conditions under which Einstein's ideas have their most dramatic effects: at speeds approaching the speed of light and in the extreme gravity that exists near black holes. For now, we'll focus only on what you actually observe on your trip, saving the *why* that lies behind your observations for the chapters that follow.

CHOOSING YOUR BLACK HOLE

If you're going to visit a black hole, the first step is to find one. You might think that would be difficult, since the term *black hole* suggests something that would be invisible against the blackness of space. There's some truth to that. By definition, a black hole is an object from which no light can escape, which means that an isolated black hole would indeed be invisibly black. However, as far as we know, all black holes are also quite massive—at least a few times the mass of our Sun, sometimes much more. As a result, we can in principle detect them by virtue of their gravitational influence on their surroundings.

A black hole's gravitational influence can reveal its presence in two basic ways. First, the black hole may be revealed by its effect on orbiting companions that are easier to see. For example, suppose you observe a star that is clearly orbiting another massive object, but the other object is not shining like it would be if it were itself a star. Since *something* must be there to explain the visible star's orbit, it's at least possible that the something is a black hole.

Second, a black hole's presence may be revealed through the light emitted by gas that surrounds it. Although we often think of space as being empty, it is not a complete vacuum; you'll always find at least a few stray atoms even in the depths of interstellar space, and the beautiful *nebulae* that you see in astronomical photos are actually vast clouds of gas. Any gas that happens to be near a black hole will end up orbiting around it, and because a black hole is both very small in size and very large in mass, the gas that is closest to it must orbit at very high speed. Gas moving at high speed tends to have a very high temperature, and high-temperature gases emit high-energy light, such as ultraviolet and X-ray light. Therefore, if you see X-ray emission coming from the region surrounding a very compact object, there's a chance that the object is a black hole.

You can see how both ideas work together in the case of the famous black hole in *Cygnus X-1*, which gets its name because it is located in the constellation Cygnus and is a source of strong X-ray emission. Cygnus X-1 is a binary system, meaning a system in which two massive objects are orbiting each other. Most binary systems have two stars orbiting each other, but in the case of Cygnus X-1, only one star can be seen. The orbit of this star tells us that the second object must have a mass that is about 15 times the mass of our Sun, yet it does not show up directly in any way. Moreover, the visible star is not hot enough to produce the X-ray emission that we observe from the system, so the X-rays must be coming from very hot gas around the second object. We thereby have both of the key clues to the possible presence of a black hole: a star orbiting a massive but unseen object, and X-ray emission suggesting that the unseen object is compact enough in size to have very hot gas orbiting it. Of course, before we conclude that the unseen object is a black hole, we must rule out the possibility that it might be some other type of small but massive object. We'll discuss how we do this in chapter 7, but current evidence strongly suggests that Cygnus X-1 really does contain a black hole.

Many similar systems are now known, and by combining observations with our current understanding of stellar lives, we've learned that most black holes are the remains of high-mass stars (stars at least 10 or so times as massive as the Sun) that have died, meaning that they have exhausted the fuel that keeps them shining during the time when they are "living" stars. With our current technology, we can identify only the black holes that, like the one in Cygnus X-1, orbit in binary systems with still-living stars. Other black holes, including those that were once single stars and those in binary systems in which both stars are long dead, are much more difficult to detect, because there is no living star with an orbit that we can observe and because the amount of gas around them is too small to lead to much X-ray emission. These black holes must be far more numerous than the ones we can detect at present, though we'll presume that we will have found them by the time you are ready for your voyage to a black hole.

In addition to black holes that are the remains of individual stars that have died, there is a second and far more spectacular general class of black holes: *supermassive* black holes that reside in the centers of galaxies (or, in some cases, in the centers of dense clusters of stars). The origins of these black holes are still mysterious, but their enormous masses make them relatively easy to identify. In the center of our own Milky Way Galaxy, for example, we observe stars orbiting a central object at such high speed that the object must have a mass about 4 million times that of the Sun, yet its diameter is not much larger than the diameter of our solar system. Only a black hole can account for so much mass being packed into such a small volume of space. Most other galaxies also appear to have supermassive black holes in their centers. In the most extreme cases, these black holes have masses that are *billions* of times the Sun's mass.

With this general background of black hole locations, we are ready to choose a target for your trip. We could in principle choose any black hole, but your trip will work best if we select one that is relatively nearby and that does not have much hot gas around it to interfere with our experiments. Although we have not yet identified such a black hole, statistically there's a decent chance that one exists within about 25 light-years of Earth. Let's therefore assume that your imaginary voyage will take you to a black hole that is 25 light-years away.

THE ROUND-TRIP FROM EARTH

In *Star Trek*, *Star Wars*, and much other science fiction, a 25-light-year trip is little more than a jaunt around the corner, and it's true that it's practically next door in terms of the galaxy as a whole. You can see why by looking at the painting of our Milky Way Galaxy in figure 1.1. Our galaxy is about 100,000 light-years across, with our solar system located about halfway out from the center to one of the edges. Because the 25 light-years from Earth to our black hole is only 0.025% of the 100,000-light-year galactic diameter, you can cover the entire length of our 25-light-year trip just by touching a sharp pencil point to the painting.

Approximate location
of our solar system

FIGURE 1.1 The Milky Way Galaxy

This painting represents our Milky Way Galaxy, which is about 100,000 light-years in diameter. If you touch a sharp pencil point to the painting, the tip will cover a length representing much more than the 25 light-years of our imaginary trip to a black hole.

Nevertheless, 25 light-years is a pretty long way in human terms. A light-year is the *distance* that light can travel in one year, and light travels really, really fast. The speed of light is about 300,000 kilometers per second (186,000 miles per second), which means that light could circle Earth nearly eight times in a single second. If you multiply the speed of light by the 60 seconds in a minute, 60 minutes in an hour, 24 hours in a day, and 365 days in a year, you'll find that a light-year is just a little less than 10 trillion kilometers (6 trillion miles), so the 25-light-year distance means a trip of nearly 250 trillion kilometers in each direction.

There are a number of ways to try to put this distance in perspective. My personal favorite is to start by visualizing our solar system at one *ten-billionth* of its actual size, which is the size at which it is depicted in the Voyage Scale Model Solar System (figure 1.2). The Sun is about the size of a large grapefruit on this scale, while Earth is smaller than the ball point in a pen and located about 15 meters from the Sun. The Moon—which is the farthest a human being has ever traveled—is only about a thumb-width from Earth on this scale. If you visit one of the Voyage models, you can walk from the Sun to Earth in about 15 seconds, and it only takes a few minutes to reach the outermost planets. But a light-year on this scale is about 1,000 kilometers (because a light-year is 10 trillion kilometers, and 10 trillion divided by 10 billion is 1,000), which means you'd have to walk across the United States (about 4,000 kilometers) just to go the 4-light-year distance of the nearest stars on this scale. As for our black hole, its 25-light-year distance means we couldn't even fit it on Earth using the Voyage model scale.

This tremendous distance presents the primary challenge of the trip. No current technology could get you to the black hole within our lifetimes, or for that matter within many lifetimes. The fastest spacecraft we've ever built are traveling out into space at speeds of around 50,000 kilometers per hour, which is the same as about 14 kilometers per second. This is quite fast by human standards; in fact, it is about 100 times as fast as a "speeding bullet." However, it is less than 1/20,000 of the speed of light, which means that at this speed, it would take more than 20,000 years to go the distance that light

FIGURE 1.2 This photo shows the Sun (the sphere on the nearest pedestal) and inner planets in the Voyage Scale Model Solar System in Washington, D.C. Voyage depicts our solar system on a scale of 1 to 10 billion. While you can walk to all the planets in a few minutes on this scale, you'd have to cross the United States to reach the nearest stars, and the 25 light-year distance to our black hole would not fit on Earth with this scale. Similar Voyage models are now open in numerous cities; visit www.voyagesolarsytem. org for more information.

travels in a single year. For a spacecraft that takes more than 20,000 years to travel a single light-year, the 25-light-year trip to the black hole would take more than 500,000 years in each direction.

Your imaginary trip is therefore going to require imaginary technology as well. Your first choice probably would be to imagine something like *Star Trek*'s warp drive, which might allow you to make the trip in weeks or less, but I'm going to quash that wish for now. While warp drive or something similar *might* be possible (an idea we'll discuss later), such things fall outside the realm of what we can currently understand

and verify in science. Therefore, your speed will be limited by Einstein's prohibition against faster-than-light travel. Einstein's theory does not, however, limit how close you can get to the speed of light. That is limited only by the practicalities of finding a way to get to very high speed. So let's suppose that future engineering will find some way to make very-high-speed travel possible, allowing you to make the trip at a speed that is 99% of the speed of light. To simplify our notation, we'll call this 0.99c, where the letter c represents the speed of light.

It's easy to see how the trip will play out from the point of view of your friends who stay home on Earth. Because you'll be traveling only slightly slower than light, it will take you only slightly longer than light to make the trip. More precisely, because light makes the trip in 25 years in each direction, or 50 years for the round-trip, your round-trip at 0.99c will take 50 years and 6 months (which is 50 years divided by 0.99). If you leave Earth early in the year 2040, and allow for 6 months of experiments at the black hole, you'll return home early in the year 2091.

By our ordinary intuition, we'd expect the trip to seem much the same for you, taking 51 years, including your time at the black hole. *But it wouldn't.* Here's what would really happen.

To keep things simple, let's suppose that you make the entire trip at the speed of 0.99c. (In reality, you'd be crushed by the forces associated with suddenly accelerating to such high speed—and with decelerating at your destinations—but we'll ignore that.) As soon as you get going, you'll find your first big surprise: Stars in the vicinity of the black hole will suddenly be much brighter than they were previously,[1] as if they were suddenly much closer to you. Indeed, if you could measure it,

---

1. At high speeds, what you actually would *see* with your eyes is more complex than I've let on here, because several additional factors besides the change in distance come into play. In this case, for example, the appearance of stars would be affected not only by the shortened distance but also by the Doppler effect caused by your motion toward (or away from) them, and you'll also see optical effects caused by the differences in light-travel time for the light from objects at different distances. Throughout this book, when I say "see," what I really mean is what you would *infer* after taking into account all the factors that affect actual appearances.

you'd find that the distance to the black hole was no longer the 25 light-years that you had measured while on Earth, but instead had shrunk to about 3½ light-years. As a result, your speed of 0.99c would allow you to reach the black hole in only a little more than 3½ years. The return trip would take the same, so along with your 6 months at the black hole, you'd be gone from Earth for a total of about 7½ years. If you marked the days on a calendar after leaving early in 2040, your calendar would say it is mid-2047 when you returned home to Earth.

Stop and think about this for a moment. Your calendar would say that you were gone for only 7½ years, to the year 2047. You would have needed only 7½ years' worth of supplies for the trip, and you would be only 7½ years older than you were when you left. But the calendars of everyone who had remained on Earth would say that it is now the year 2091. Your friends and family would be 51 years older than they were when you left. Society would have gone through 51 years' worth of cultural and technological changes. In other words, you would return home to find that 51 years had passed on Earth, even though only 7½ years had passed for you. You would have felt nothing unusual during your trip, yet *time itself* would have passed more slowly for you than it had on Earth.

If you have not studied Einstein's theories previously, you may be having a hard time believing me. That's OK, since I haven't yet given you reason to believe me; I hope to do that in the coming chapters. For the moment, suffice it to say that you've just seen an example of the dramatic effects that Einstein's theory predicts for travel at speeds near the speed of light. Now, let's return to the midpoint of your trip, as you approach the black hole.

ENTERING ORBIT

The first step in understanding your approach to the black hole is to remember how travel through space is different from travel on Earth. On Earth, turning off the engines on your vehicle will ultimately cause you to slow down and stop. The reason is friction, either with the ground, with water, or with air. In space, where there's no friction, you can turn off your engines and still keep going forever, as long as you don't crash into anything. Aside from firing your engines, the only thing that will affect your speed and trajectory is gravity. Therefore, to understand what will happen as you approach the black hole, we need to understand how its gravity can affect your trajectory.

In daily life, we usually think of an orbit as a path that goes round and round. When we are speaking of space, however, an orbit is *any* path that is governed solely by gravity, and it does not matter whether the source of the gravity is a planet, a star, a black hole, or anything else.

The general properties of orbits were first worked out more than 300 years ago by Isaac Newton. He discovered that orbits can have three basic shapes: ellipses, parabolas, and hyperbolas. (Circles are counted as a special type of ellipse in much the same way that squares are a special type of rectangle.) These shapes are often called "conic sections," because you can make them by slicing a cone at different angles. Figure 1.3 shows the three shapes and how you make them.

You should pay special attention to three key points as you look at the allowed orbits in figure 1.3. First, notice that ellipses (including circles) are the only orbital paths that go "round and round," returning to the same place with each orbit. This explains why we usually think of ellipses when we think of the word *orbit*, because it means that all moons have elliptical orbits around their planet, all planets have elliptical orbits around their star, and all stars have elliptical orbits around their galaxy.

The second key idea to notice in figure 1.3 is the difference between *bound* and *unbound* orbits. We say that ellipses are *bound* orbits because objects on them remain bonded to the central object by gravity. Parabolas and hyperbolas are *unbound* because objects following them come in and swing past the central object just once, never to return again, meaning that the central object's gravity does not have a

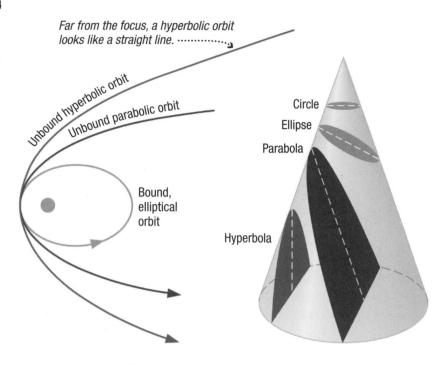

Far from the focus, a hyperbolic orbit looks like a straight line. ·················⟩

Unbound hyperbolic orbit

Unbound parabolic orbit

Bound, elliptical orbit

Circle

Ellipse

Parabola

Hyperbola

FIGURE 1.3 Orbits Allowed by Gravity

More than 300 years ago, Isaac Newton discovered that gravity allows only the three basic orbital paths shown on the left. The cone on the right shows how these shapes can be made by slicing through a cone, which is why they are called *conic sections*; note that a circle is simply a special case of an ellipse.

permanent hold on them. Objects coming in from afar, such as space-craft (and many comets), are clearly not on bound elliptical orbits, so they must be on either parabolic or hyperbolic orbits. Most unbound orbits are hyperbolic.[2]

2. The major difference between a parabola and a hyperbola is its shape at great distance from the central object. A parabola is curved everywhere along it, while at large distances a hyperbola is indistinguishable from a straight line. Mathematically, there is a much wider range of possibilities for hyperbolas than parabolas (which are the borderline case between hyperbolas and ellipses), which is why most unbound orbits are hyperbolic.

The third key idea is that ellipses, parabolas, and hyperbolas represent a complete list of allowed orbits. Because *sucking* is not on the list, you now understand the fact I told you earlier: *Black holes don't suck*. Gravity is gravity, determined by an object's mass. From a distance, the gravity of a black hole is no different than the gravity of any other object of the same mass. It is only when you get very close[3] to a black hole that you begin to notice effects different from those of gravity as understood by Newton; for the time being, we'll assume you remain at a distance at which the basic orbital paths are those discovered by Newton.

We can now apply these ideas to your spacecraft. Because you are coming in toward the black hole from afar, you are on an unbound, hyperbolic orbit. Therefore, unless you fire your engines, you will continue on this unbound path, which means you will simply swing past the black hole once, never to return to it again. The only way around this would be if you happen to be aimed nearly dead-on at the black hole, in which case you'd plunge into it before your orbit took you back out. However, this is extremely unlikely. Remember that although black holes are large in mass, they are small in size. For example, a black hole with a mass 10 times the Sun's mass is only about 60 kilometers across in size. That makes it a target barely as large as a big city on Earth, and smaller than many asteroids. After traveling 250 trillion kilometers from Earth, you'd have to be the unluckiest person in history to be aimed directly at the black hole by accident.

The practical consequence of all this is that the only way to prevent yourself from flying past the black hole at high speed is to use your engines to slow your spaceship down. If you fire them just right, you can put your spacecraft on a bound orbit around the black hole.

---

3. "Very close" in this case means within about 100 kilometers of the black hole. (More generally, it means within about two *Schwarzschild radii* of the event horizon, an idea we'll define in chapter 7.) It is only within this distance that you could no longer follow stable Newtonian orbits. In fact, it might even seem like you are being "sucked in" within this region, though in chapter 7 we'll see that there's a better way to think about it.

Let's assume that you've done this, so that your spaceship is now "parked" with its engines off in a circular orbit a few thousand kilometers away from the black hole. The force of gravity holding your spaceship in orbit will be quite strong here; the strength of gravity depends on both the central object's mass and your distance from it, and a few thousand kilometers is quite close to an object with a mass greater than that of the Sun. Nevertheless, you are perfectly safe—there's no need to worry about being "sucked in"—and you can orbit round and round in this location for as long as you wish.

OBSERVATIONS FROM ORBIT

From your orbital vantage point, things will at first seem quite mundane. Unless your spaceship is rotating, you'll be floating weightlessly within it, and any clocks on the wall will be recording time normally. At a distance of thousands of kilometers, and with very little gas around it to emit light, the black hole will be almost completely invisible. Aside from the fact that you'll be orbiting an unseen object at a relatively high speed (because strong gravity means a stable orbit requires high speed), there will be little to indicate that you're in the vicinity of a black hole.

Nevertheless, you didn't come all this way for nothing, so you decide to conduct some experiments. For your first experiment, you pick two identical clocks out of the ship's supply closet, each of which has numerals that glow with blue light. Setting them both to the same time, you keep one clock aboard the ship, and push the other one gently out the air lock toward the black hole, with a small rocket attached to it. You've pre-programmed the rocket to fire continuously in such a way that the clock falls only slowly away from your ship and toward the black hole. You'll quickly begin to notice the outside clock behaving strangely.

Although both clocks start out reading the same time in their blue numerals, you'll soon observe the clock falling toward the black hole to be running noticeably slow. Moreover, as it drops away from you, its blue numerals will gradually change color, becoming increasingly red. Both observations—the slower ticking of the clock and its reddening numerals—are consequences of a key effect predicted by Einstein: *Time runs slower in stronger gravity.*

It's fairly obvious that slower time would make the falling clock tick more slowly. The reason for the reddening numerals is a bit less obvious, but you can understand it as follows. The clock itself would "feel" normal (if it had feelings), so from its own point of view, its numerals would be emitting blue light just like normal. All forms of light can be thought of as waves that vibrate with some frequency; the frequency of blue light happens to be about 750 trillion cycles per second, while the frequency of red light is somewhat lower (about 400 trillion cycles per second). Now, remember that from your vantage point on the ship, time on the falling clock is running slow, which means that one second for the clock is *longer* than one second for you. Therefore, during each of your seconds, you see only a portion of the 750 trillion cycles that the clock emits during each of its own seconds. You'll therefore observe the frequency of the emitted light to be lower than the 750 trillion cycles per second of blue light, and lower frequency means redder color. This effect, in which objects in strong gravity emit redder light than they would otherwise, is called a *gravitational redshift.*

Let's go back to watching the clock. To keep the clock falling only slowly toward the black hole, the clock's rocket will have to fire ever more powerfully as it descends away from your ship. This is unsustainable, because the rocket fuel will run out. When it does, it will be like the floor has been pulled out from under it, and the clock will begin to accelerate rapidly toward the black hole. Here's where things really get strange.

From the point of view of the clock, it is falling toward the black hole in the same way that a rock falls toward Earth, except that the gravity is far stronger. Therefore, the clock will accelerate to higher and higher speeds as it approaches the black hole, which means it will quickly fall into the black hole. Be sure to note that it has *fallen* into the black hole just like a rock falls to the ground on Earth; it has not been "sucked in."

The clock's viewpoint is simple enough, but things look quite different from your vantage point on the ship. At first, you'll see the clock accelerating toward the black hole, much as the clock would see itself doing. But as you watch the clock get closer to the black hole, its acceleration will be offset by the slowing of time. The ticking of the clock will continue to become slower and slower as it approaches the place known as the *event horizon* of the black hole. In fact, if you could continue to watch the clock, you'd see time on it come to a halt as it reached the event horizon, which means it would *never* actually fall past that point.

However, you won't actually be able to see the clock's face become frozen in time, because of the gravitational redshift. The same effect that made the clock's numerals shift from blue to red in color will continue, so as the clock falls the frequency of its light will get lower and lower. Light with frequency lower than that of visible light is what we call *infrared light*, and light of even lower frequency makes what we call *radio waves*. You might therefore be able to watch the clock for a short time with an infrared camera, and after that with a radio telescope, but before the clock reaches the event horizon, its light will have reached such low frequencies that no conceivable telescope could detect it. It will vanish from your view, even as you realize that time is about to come to a stop on it.

PLUNGING IN

Back on the spaceship, you and your crewmates are busy discussing what you've just seen, when curiosity overwhelms the better judgment of one of your colleagues. He slips away from the conversation, hurriedly climbs into a space suit, grabs the other clock, and jumps out of the air lock on a trajectory aimed straight for the black hole. Down he falls, clock in hand. (For reasons we'll discuss shortly, he will die long before he reaches the black hole. But let's ignore that for the moment, and imagine that he could still observe as he fell.)

He watches the clock as he falls, but because he and the clock are traveling together, its time seems to run normally and its numerals stay blue. That is, although from the spaceship you'd see the clock running slow and its numerals becoming redder in color, he would notice nothing unusual about it. He would notice something strange only when he looked back up at the spaceship. For example, if he could fire his space suit's rockets strongly enough to momentarily stop his fall, and he then looked back up,[4] he would see your time running *fast*, and your light shifted to a *bluer* color—the opposite of what you see happening to him. When his fuel ran out, the black hole's enormous gravity would make his acceleration toward it resume instantly. In fact, because gravity becomes stronger as you get closer to a massive object, your colleague's acceleration would become greater as he fell toward the black hole, which means that his speed would grow faster at a rate that was both enormous and increasing. In a fraction of a second, he would plunge past the event horizon, becoming the first human being ever to fall into a black hole.

You might wonder what he sees once he is inside the black hole, but don't hold your breath waiting for his report back. Remember, from your point of view on the spaceship, he will *never* cross the event horizon. You'll see time coming to a stop for him just as he vanishes from view due to the gravitational redshift of light. That brings us to some good news and some bad news.

The good news is that when you return home, you can play a video for the judges at your trial, proving that your crewmate is still outside the black hole. They can hardly convict you of complicity in his plunge into the black hole if he still hasn't quite reached it. The bad news is that despite still being outside the black hole, your colleague is dead. In fact, it would have been a fairly gruesome (but rapid) death,

---

4. I've had your colleague momentarily stop his fall to look back up so that you can see the symmetry of the situation: You observe his time running slower and his light redshifted, while he observes your time running faster and your light blueshifted. At all other times during his fall, he will actually see you redshifted because of his high velocity away from you.

caused by an unavoidable side effect of getting too close to the black hole. This side effect comes about for the same reason that we have tides on Earth.

Tides on Earth arise primarily from the Moon's gravitational influence and the fact that our planet is about 13,000 kilometers in diameter, which means that whatever side is facing the Moon at a particular moment is about 13,000 kilometers closer to the Moon than the other side. Because the strength of gravity depends on distance, the Moon exerts a stronger gravitational pull on the parts of Earth that are nearer to it. This difference in the Moon's gravitational pull on different parts of our planet effectively creates a "stretching force" that makes our planet slightly stretched out along the line of sight to the Moon and slightly compressed along a line perpendicular to that. You can see a similar effect by pulling both ends of a rubber band in the same direction (just as the Moon's gravity pulls all parts of Earth in the same direction), but pulling one side more than the other (just as gravity pulls harder on the side facing the Moon). The rubber band will be stretched along its length and squeezed along its width, despite the fact that both ends are moving in the same direction.

The tidal stretching caused by the Moon's gravity affects our entire planet, including both land and water, inside and out. However, the rigidity of rock means that land rises and falls with the tides by a much smaller amount than water, which is why we notice only the ocean tides. The stretching also explains why there are generally *two* high tides (and two low tides) each day: Because Earth is stretched much like a rubber band, the oceans bulge out both on the side facing toward the Moon and on the side facing away from the Moon. As Earth rotates, we are carried through both of these tidal bulges each day, so we have high tide when we are in each of the two bulges and low tide at the midpoints in between.

More generally, a *tidal force* is simply the difference between the tug of gravity on one side of an object and the tug on the other side.

The strength of the tidal force therefore depends on two things: (1) the distance across the object and (2) the strength of the gravity acting on it. The first explains why the Moon's tidal force has no effect on our bodies even though it has a noticeable effect on our planet: The distance from our heads to our toes is far too small for the Moon's relatively weak gravity to create a measureable tidal force. But if you come close enough to a black hole, its immense gravity creates a tidal force trillions of times stronger than that of the Moon, so that even over the relatively short distance from your crewmate's head to his toes—or, for that matter, across his waist if he turned sideways—he'd feel himself being stretched so hard that he would be ripped apart. Sadly, it would be only his blood and guts that would experience the inside of the black hole.

You may be wondering if there's any way that a human could avoid a gruesome death and learn what is inside a black hole. For black holes that are the remains of individual stars, like the one we've just visited, the answer is probably no, because there is no practical way to counteract a tidal force. However, you could in principle survive the plunge across the event horizon of a supermassive black hole. Although you could no more escape from the event horizon of a supermassive black hole than from that of any other black hole, the larger size of a supermassive black hole would make its tidal force at the event horizon much weaker. You could therefore live, at least for a little while, to see the inside of the black hole.

It's interesting to think about what you'd see on your way in. Remember that, from the point of view of those of us outside the black hole, including everyone at home on Earth, it would take you *forever* to fall into the black hole. There would be no point in us waiting for you to enter and report back, since you'd never get there. From your point of view, however, you'd plunge into the black hole at very high speed. In principle, the symmetry of the situation means that a lot of time passes on Earth as you approach the event horizon, and you might therefore guess that you could watch Earth's future history go by just by looking back before you crossed the event horizon. Unfortunately, that's not the case, because of the way light from Earth would be distorted due both to your speed and to the black hole's gravity. But even if the light was not distorted, you still couldn't watch future stock markets and then go home to invest, because as far as we know, there is no escape from a black hole. The event horizon is the point at which the speed required to leave the vicinity of the black hole (the escape velocity) is the speed of light, and because Einstein's theory tells us that no material object can reach that speed, nothing can escape from within the event horizon. Presumably, you would simply continue to fall toward the black hole's center, or *singularity*, meeting your tidal demise some time before you arrive.

SCIENCE AND SCIENCE FICTION

Wait, you say—science fiction writers and some scientists have suggested that there might be a way to survive a trip through a black hole, perhaps even using the black hole as a "wormhole" to travel between two distant points in the universe. It's a nice idea, but the key word here is "might." Although there does not seem to be anything in the known laws of physics to prohibit such an idea,[5] there's also nothing that says it is correct.

This brings us to an important point about the nature of science. Science is about *evidence*. The reason we can describe the strange effects on time that you experience during your voyage to the black hole is *not* because a smart guy named Einstein thought them up. Rather, it is because scientists have carefully tested the predictions that Einstein made, and these tests have been conducted for a wide range of conditions. While we do not have the technology to test conditions quite

---

5. In fact, exact mathematical solutions for rotating black holes contain one-way wormholes that connect to "other universes." However, the solutions also indicate that these wormholes are unstable and could not be used for physical travel.

as extreme as those at the event horizon of a black hole, every test to date suggests that Einstein was correct. Without these tests, Einstein's ideas would be nothing more than speculation.

In essence, evidence is the difference between science and science fiction. Science fiction is free to imagine anything that doesn't violate known laws (and sometimes even things that do), which leaves open an almost limitless number of possibilities. In contrast, science is limited to the relatively narrow range of ideas that we can currently test, or to exploring ideas that we may be able to test in the future.

Although this basic difference between science and science fiction is probably clear, it often creates confusion at the edge of current knowledge. Take ideas about the insides of black holes. Physicists can use the known laws of nature to predict what conditions would be like after you cross the event horizon. Indeed, I've done just that in assuming that you would continue to fall toward singularity until being killed by tidal forces. Since these predictions are based on tested ideas, it's tempting to assume that they must be correct. However, because we do not yet know of any way to test our predictions about what occurs inside a black hole, even the most solid-seeming predictions remain speculative. The level of speculation is even greater for ideas for faster-than-light travel, such as hyperspace, wormholes, or warp drive. It's possible that some of these ideas will someday prove to be valid, but until we can test them, they're more science fiction than science.

In this book we will focus on the evidence-based science of Einstein's ideas. We will stay away from science fiction, and even from more straightforward mathematical extrapolations, except for a few brief mentions about those ideas that you may have heard in popular culture. This stick-to-the-evidence approach will make this book somewhat different from most other popular science books, since the demands of the market usually make authors focus on the more speculative boundaries between science and science fiction. The advantage of our evidentiary approach is that everything in this book is established science. In fact, much of what we'll discuss has been known to scientists for more than a century, since Einstein published the first part of his theory of relativity in 1905. Nevertheless, old news is not always boring news, and if you haven't studied it before, I think you'll find Einstein's ideas not only mind-blowing but also *important*, in the sense that they will change the way you view the universe.

Part 2

EINSTEIN'S SPECIAL THEORY OF RELATIVITY

2

RACING LIGHT

ALTHOUGH WE usually speak of *the* theory of relativity, Einstein actually published the theory in two distinct parts. The first part, known as the *special theory of relativity*, was published in 1905. This is the theory that explains the slowing of time that made you age less than those who stayed home on Earth during your voyage to the black hole. It's also the theory that tells us that nothing can travel faster than light, and from which Einstein discovered his famous equation $E = mc^2$. You may be thinking, "Well, that *is* pretty special, but isn't it odd to put the word *special* in its name?" Yes, it would be odd. The real reason for the word *special* is to distinguish this theory from a theory he published a decade later, which we call the *general theory of relativity*.

As the names imply, the special theory of relativity is essentially a subset of the general theory. In particular, the special theory applies only to the special case in which we ignore any effects of gravity, while the general theory includes gravity. Hence it is the general theory that explains your observations in the strong gravity of the black hole; it is also the theory through which we understand the structure of the universe as a whole, including the observed expansion of the universe.

For much the same reason that Einstein worked out the special theory before the general one, it is easier to learn about relativity by starting with the special case. We'll therefore proceed in this way, beginning with one of the most famous claims of the special theory.

IT'S THE LAW

Perhaps you've seen T-shirts or posters that read: "300,000 Kilometers per Second: It's Not Just a Good Idea, It's the Law!" They're quite popular among aspiring physicists, though their message is arguably the least popular claim that comes from Einstein's special theory of relativity. The reason for this idea's unpopularity is probably obvious: No one likes to be told what they cannot do. Nevertheless, Einstein's prohibition on faster-than-light travel rests on such a solid foundation of evidence that even science fiction writers generally avoid violating it. The spaceships in *Star Trek* and *Star Wars* never actually travel *through* space at speeds greater than the speed of light; instead, they somehow bend or warp space (as in *Star Trek*) to bring faraway points closer together, or temporarily leave space altogether (as in *Star Wars*) so that they can jump through hyperspace and emerge in some other location. Even if such "loopholes" in the laws of nature someday prove really to exist, they won't change the basic fact that you cannot hop into a spaceship and accelerate to a speed faster than light.

Why not? Most people have heard of this law, but most also assume that there's got to be some way around it. After all, there've been many cases in history where people have done what was once thought impossible. Famous examples include claims by well-respected scientists in the twentieth century that the sound barrier could never be broken, or that we could never land people on the Moon. But if relativity is correct—and the evidence strongly indicates that it is—then light speed is different. The problem is that the speed of light is not a barrier to be broken like the speed of sound, or a challenge like reaching the Moon. We always knew that there were things that go faster than sound, and we always knew that there were objects that could reach the Moon. The question was whether *we* could do it.

Because it can be so hard to accept, I'll start by telling you the conclusion that we'll work toward in this chapter: Relativity tells us that everyone always measures the speed of light to be the same, and this agreement about the speed of light leads inevitably to the fact that you cannot outrace your own light. And if you can't outrace your own light, then others looking at you will always see you moving slower than any light you emit or reflect. In some sense, what relativity really tells us is that the speed of light is a fundamental property of nature, in much the same way that the existence of a North Pole is a fundamental property of a rotating planet. Asking how you can go faster than light is somewhat like asking how you can walk north from the North Pole (from which *all* directions are south). It's a question that doesn't really make any sense, at least once you understand the meaning of the North Pole—or of the speed of light.

Before we proceed, there are two important caveats that you should be aware of. First, when we speak of the speed of light in relativity, we mean the speed that light travels through empty space. This speed, which is 300,000 kilometers per second, is actually the *maximum* speed of light; light travels slower when it passes through materials such as water, air, or glass, and in recent years scientists have found ways to slow light down to pedestrian speeds in the laboratory. Obviously, it is possible to outrace light that is moving unusually slowly. What you can't outrace is light traveling freely through space.

The second caveat is that the common statement "nothing can go faster than light" is not quite what relativity really tells us, which is better stated as *nothing can outrace light*.[1] To take a specific example, modern astronomy suggests that there are probably galaxies that lie hundreds of billions of light-years away from us—far beyond the bounds of our *observable universe*, meaning the portion of the universe that we can in principle see—that are being carried away from us with the expansion of the universe at speeds far faster than the speed of light. This does not violate the special theory of relativity, because

---

1. Mathematically, the theory of relativity actually allows for particles called *tachyons* that *always* travel faster than light; that is, just as we could never go faster than light, tachyons could never go slower than light. Most physicists doubt that tachyons really exist, and even if they do, it would not change the fact that nothing *that starts out at a speed slower than light* can ever accelerate to a point at which it would outrace light.

the separation of these galaxies from us does not involve anyone (or anything) outracing any light. If we tried to travel to one of these galaxies, we'd never be able to get there; our light could not catch up to them, so neither could we. An equivalent way to think about it is to say that the prohibition on faster-than-light travel applies only to the ability to transmit matter or information from one place to another, or to say that nothing can travel *through space* at a speed faster than light. That's the way you'll hear it described in most relativity texts, though I find it easier just to remember that "nothing can outrace light."

WHAT'S RELATIVE IN RELATIVITY?

The first step in understanding relativity is to be clear about exactly what is relative. Contrary to a common belief, Einstein's theory does *not* tell us that "everything is relative." Rather, the special theory of relativity takes its name from the idea that *motion* is always relative.

The idea that motion is relative may at first seem counterintuitive. After all, if you watch a car driving by on the highway or an airplane flying overhead, it may seem obvious that it is the car or the airplane in motion, while you remain stationary on the ground. But, in fact, it's not quite as obvious as it seems. To understand why, imagine a supersonic airplane flying from Nairobi, Kenya, to Quito, Ecuador, at a speed of 1,670 kilometers per hour (about 1,040 miles per hour). Now answer this question: How fast is the plane going?

At first, the question sounds trivial, since I've just told you that the plane is going 1,670 kilometers per hour. But wait. Nairobi and Quito are both nearly on Earth's equator, and the equatorial speed of Earth's rotation happens to be the same 1,670 kilometers per hour, but in the opposite direction (figure 2.1). Therefore, if you viewed the flight from the Moon, the plane would appear to stay put while Earth rotated beneath it. When the flight began, you would see the plane lift off the ground in Nairobi. The plane would then remain stationary while Earth's rotation carried Nairobi away from it and Quito toward it.

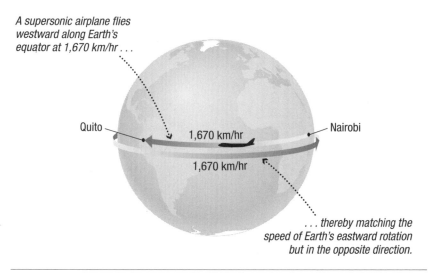

A supersonic airplane flies
westward along Earth's
equator at 1,670 km/hr . . .

Quito

1,670 km/hr

1,670 km/hr

Nairobi

. . . thereby matching the
speed of Earth's eastward rotation
but in the opposite direction.

FIGURE 2.1 Imagine a supersonic airplane flying westward from Nairobi to Quito at a speed of 1,670 kilometers per hour. This speed happens to precisely match Earth's equatorial rotation speed, which goes in the opposite direction. So how fast is the airplane *really* moving?

When Quito finally reached the plane's position, the plane would drop back down to the ground.

So what is the airplane *really* doing? Is it flying at 1,670 kilometers per hour, or is it staying still (zero speed) while Earth moves beneath it? According to the theory of relativity, there is no absolute answer to this question. You can describe motion only if you state what the motion is relative to. In other words, you can validly say that "the airplane is moving relative to Earth's surface at 1,670 kilometers per hour," and you can validly say that "viewed from the Moon, the plane appears stationary while Earth rotates beneath it." Neither viewpoint is any more or less valid than the other.

In fact, there are many other equally valid viewpoints about the plane's flight. Observers looking at the solar system from another star would see the plane moving at a speed of more than 100,000 kilometers per hour, because that is Earth's speed in its orbit around the Sun. Observers living in another galaxy would see the plane moving with the rotation of the Milky Way Galaxy at about 800,000 kilometers per hour.

In the language of relativity, we say that any description of the airplane's motion depends on the observer's *frame of reference*. Each of the different viewpoints on the airplane's motion—such as the viewpoints from Earth's surface, from the Moon, from another star, and from another galaxy—represents a different reference frame. More generally, we say that two objects (or people) share the same frame of reference only if they are stationary relative to each other.

THE ABSOLUTES OF RELATIVITY

The name "theory of relativity" is in some sense a good name, in that the relativity of motion is a fundamental part of the theory. But in another sense it is a misnomer, because the foundations of the theory actually rest on the idea that two particular things in the universe are absolute:

1. The laws of nature are the same for everyone.
2. The speed of light is the same for everyone.

Every astounding idea that comes from Einstein's special theory of relativity—including the strange ways in which time and space are different for you during your voyage to the black hole than they are for people on Earth—follows directly from these two seemingly innocuous absolutes. For that reason, let's briefly explore what these two absolutes really mean.

The first absolute, that the laws of nature are the same for everyone, probably is not surprising. Indeed, this idea predates Einstein by centuries, going back to Galileo. For example, this idea explains why you feel no sensation of motion when you are traveling on an airplane on a smooth flight. In this case, the only difference between your airplane reference frame and a reference frame on the ground is your relative motion over the ground. As long as that motion is at a constant velocity, so that you won't feel any forces that differ from those felt on the ground, you can perform experiments in the airplane and get the same results that you'd get if you did them in a laboratory on the ground. The fact that you would get all the same experimental results means that you would come to precisely the same conclusions about the laws of nature.

The second absolute, that the speed of light is the same for everyone, is far more surprising. In general, we expect people in different reference frames to give different answers for the speed of the same moving object. For example, suppose you are on an airplane that is traveling relative to the ground at a speed of 800 kilometers per hour. If you roll a ball down the aisle toward the front of the airplane, you'll say that the ball is moving fairly slowly. In contrast, people on the ground will say that the ball is moving past them quite fast, because they'll see it moving at the speed you roll it *plus* the airplane's 800 kilometers per hour.

Now, suppose that instead of rolling a ball, you turn on a flashlight. By the same logic that we used for the ball, you would expect that a person on the ground would say that the flashlight beam is traveling 800 kilometers per hour faster than you would say it is going inside the airplane. *But that is not the case*, because the second absolute of relativity tells us that everyone always measures the same speed of light. Therefore, no matter how precisely you measure it, both you and people on the ground will say that the light beam is going *exactly* the same speed; this speed will, of course, be the speed of light, or 300,000 kilometers per second.

The absoluteness of the speed of light is so surprising that we should take a moment to be clear about why it is so important. As I said above, every astounding consequence of special relativity follows directly from the two absolutes. Given that the first is unsurprising and was long suspected, all the consequences of relativity in essence stem from the single surprising idea that everyone always measures the same speed of light. In other words, if this idea is correct, then all of special relativity will make perfect sense. Conversely, if it is not correct, then the entire theory will fall apart.

So what makes us so confident that Einstein was right? Remember, observations and experiments are the ultimate arbiters of truth in science, and *the absoluteness of the speed of light is an experimentally verified fact*. The first clear demonstration of this fact came with an experiment performed in 1887 by A. A. Michelson and E. W. Morley. In their now-famous *Michelson-Morley experiment*, they discovered that the speed of light is not affected by Earth's motion around the Sun. Today, we can measure the absoluteness of the speed of light in many other ways. To take a simple but ubiquitous example, every star and every galaxy in the sky is moving relative to Earth at a different speed. Some distant galaxies are moving away from Earth at speeds that are close to the speed of light. Yet, if you were to measure the speed of the light coming from any of these objects, you would find that it is always the same 300,000 kilometers per second that you measure for light beams emitted on Earth. There's simply no way around it: Experiments show that you will always measure the same speed of light (through empty space), no matter how you are moving relative to the light's source.

THOUGHT EXPERIMENTS AT LOW SPEED

Proceeding much as Einstein did, we can now build the special theory of relativity by engaging in a series of *thought experiments*, in which we think through experiments that we won't actually perform, but that could be performed in principle. Keep in mind that while thought experiments are crucial in helping us understand and predict the consequences of the theory, they don't by themselves constitute evidence for it. We consider the thought experiments to be valid only because real experiments (which we'll discuss later) have borne out all the conclusions that we will draw from those we do through thought.

Einstein's own thought experiments often involved thinking about the motion of trains. However, both because relative motion is easier to visualize in space and because special relativity neglects the role of gravity, we'll do our thought experiments with spaceships. We will assume that our spaceships have their engines off, so that everything within them is weightless and floats freely. For this reason, we say that the reference frames of these spaceships are *free-float frames* (sometimes called "inertial" reference frames). In case you are wondering *why* you would be weightless in a spaceship with its engines off, here's an easy way to think about it: The concepts of up and down have meaning only when we can describe them relative to a planet's (or other object's) surface. There are no reference points for up or down when you are in deep space, so as long as your engines are off, you won't have any reason to move in any particular direction—which means you will float weightlessly.

To make sure you understand how thought experiments work, let's begin with some in which the spaceships and other objects are moving at low speeds relative to one another. Imagine that you are out in space, floating freely in your spaceship. Because you feel no forces of any kind, you will naturally consider yourself to be at rest (not moving). Now, you look out your window, and you see your friend Al floating in his own spaceship, which is moving to your right at a speed that you measure to be 90 kilometers per hour. What will Al say is happening?

Like you, Al will not feel any forces as he floats freely in his ship, and therefore he will say that *he* is the one who is stationary and that it is *you* moving (to his left) at 90 kilometers per hour. This is fine, of course; because all motion is relative, your viewpoint and Al's viewpoint are both equally valid.

Let's add a small wrinkle to the thought experiment. As Al goes by, you hop into your spacesuit, strap your feet to the outside of your spaceship (so that you won't float away), and throw a baseball in his direction at a speed of 100 kilometers per hour. What will Al say the baseball is doing? He will still consider himself to be at rest, with you moving away from him at 90 kilometers per hour. Therefore, as shown in figure 2.2, he will see the baseball coming at him at a speed of only 10 kilometers per hour.

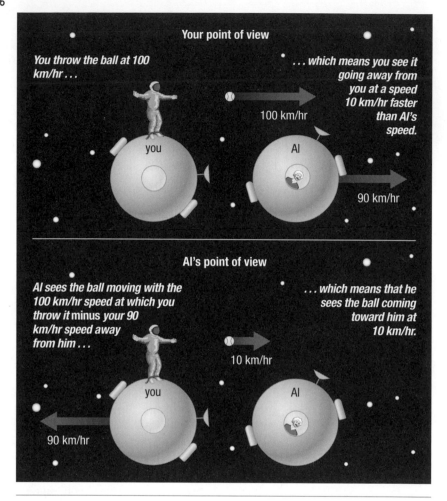

FIGURE 2.2 According to you, you are at rest, while Al is moving to your right at 90 kilometers per hour and the ball is moving in Al's direction at 100 kilometers per hour. According to Al, he is the one who is at rest, while you move away at 90 kilometers per hour and the ball comes *toward* him at 10 kilometers per hour.

We get an even more interesting result if we imagine that you throw the ball toward him at 90 kilometers per hour. Since that precisely matches the speed at which he sees you moving away from him, the ball will now be stationary in his reference frame. Think about this for a moment: Before you throw the baseball, Al sees it moving away from him at 90 kilometers per hour because it is in your hand. At the moment you release the baseball, it suddenly becomes stationary in Al's reference frame, floating in space at a fixed distance from his spaceship. Many hours later, after you have traveled far away, Al will still see the ball floating in the same place. If he wishes, he can put on his space suit and go out to retrieve it, or he can just leave it there. From his point of view, it's not going anywhere, and neither is he.

THOUGHT EXPERIMENTS AT HIGH SPEED

The absoluteness of the speed of light has not yet come into play in our thought experiments, because the speeds we have used are so small compared to the speed of light. For example, if you work it out, you'll find that the 100-kilometer-per-hour speed of the baseball is less than one *ten-millionth* of the speed of light. As you might guess, things will look a bit different when we ramp up the spaceship speeds.

Imagine that Al is now moving to your right at 90% of the speed of light, or $0.9c$. (Recall that c is the symbol for the speed of light.) As before, you and Al can both legitimately claim to be at rest. Therefore, while you say that Al is moving to your right at $0.9c$, he will say that he is the one who is stationary and that you are moving to his left at $0.9c$.

Because you consider yourself to be at rest and are floating freely in your ship, it is still easy for you to put on your space suit and strap your feet to the outside of your spaceship. This time, instead of taking a baseball, you take a flashlight. Pointing it in Al's direction, you turn it on. How will you each describe the situation?

Your point of view is easy. Al is moving to your right at $0.9c$ while the light beam is going in the same direction at the full speed of light, c. Therefore, you will say that the light beam is going $0.1c$ faster than Al, which means it will gradually catch up to him and pass him by.

Turning to Al's point of view, if we apply the same pre-relativity mind-set that we used with the baseball, we'd expect him to say that the light is coming at him at a speed of 0.1c, which is the light's speed of c minus your speed of 0.9c. But he wouldn't. The absoluteness of the speed of light demands that he measure the speed of the light beam coming toward him to be the full speed of light, not some fraction of it. In other words, the fact that you are moving away from him at 90% of the speed of light has *no effect at all* on the speed he measures for the light beam as it passes him by. Figure 2.3 summarizes this thought experiment.

YOU CAN'T OUTRACE LIGHT

What would happen in our thought experiment if instead of moving by you at 0.9c, Al were going at or faster than the speed of light? It seems like a legitimate question, but remember that if Al is going by you, he must have come from somewhere. So let's think back to where he got started; better yet, since Al says that you are the one flying by at high speed, let's make it more personal by considering how *you* got started.

You have just built the most incredible rocket ever, and you are taking it on a test ride. Soon you are going faster than anyone had ever imagined possible . . . and then you put it into second gear, and third, and so on. You keep going faster and faster and faster. When, you might wonder, will you reach the speed of light?

The key point to remember is that all motion is relative, so when we ask how fast you are going, we must also ask "according to whom?" Let's start with your own viewpoint. Imagine that you turn on your rocket's headlights. Because everyone always measures the same speed of light, you will see the headlight beams traveling away from you at the full, normal speed of light, or 300,000 kilometers per second. This will be true at all times and under all circumstances, no matter how long you have been firing your rocket engines. In other words, you cannot possibly keep up with your own headlight beams.

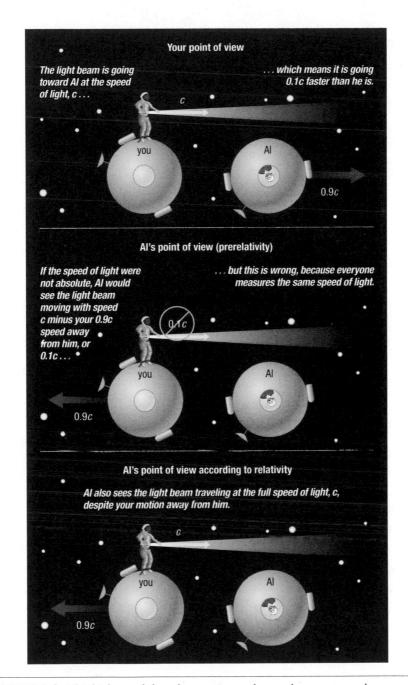

FIGURE 2.3 This high-speed thought experiment shows what we mean when we say that everyone always measures the same speed of light.

Now let's see what others will say about you. Whether it is someone back on Earth, or someone like Al in his spaceship, or anyone else, relativity tells us that there are two things that everyone will agree on: First, everyone will agree that your headlight beams are going at the speed of light, 300,000 kilometers per second. Second, because there is only one reality, everyone will also agree with you that your headlight beams are beating you (figure 2.4). There you have it: Because everyone agrees that your headlight beams are going the speed of light and that they are beating you, then everyone must inevitably conclude that you are going *slower* than the speed of light. The same argument applies to any other traveler and any other moving object, and it is true with or without headlights. All objects emit or reflect light of some type, and as long as the speed of light is absolute, no material object can ever keep up with its own light.

You can now see what I meant earlier, when I said that asking how you can go faster than light is somewhat like asking how you can walk north from the North Pole. You can't go north of the North Pole because all directions from it are south. You can't go faster than light because there's no way to catch up to light, no matter what you do. Building a spaceship to travel at the speed of light is not a mere technological challenge—it simply cannot be done.

I realize that you're probably still looking for loopholes in the logic. Maybe you're thinking about those distant galaxies I told you about, which may be carried away from us with the expansion of the universe at a speed faster than the speed of light. Doesn't that mean that you—and everyone else here on Earth—are right now traveling away from them faster than light? In a sense it does, but that's exactly why it's a moot point. If a distant object were moving away from you faster than the speed of light, its light wouldn't catch you, and your light wouldn't catch it.[2]

---

2. A caveat: As we'll discuss in chapter 8, distant galaxies aren't really "traveling" away from us with expansion; rather, expansion makes the space between us and distant galaxies grow with time. As a result, changes in the expansion rate can mean that light from galaxies being carried away from us faster than the speed of light can in some cases still end up inside the observable universe,

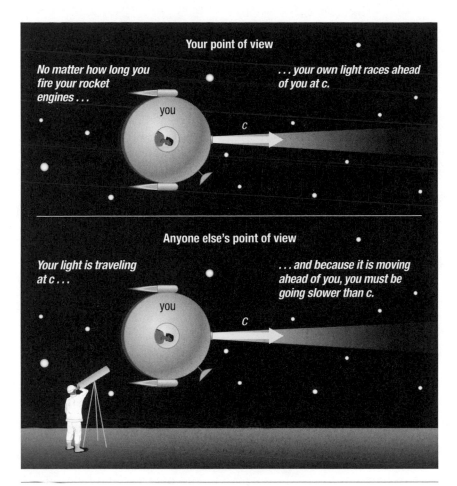

FIGURE 2.4 This illustration summarizes why the absoluteness of the speed of light means that you can never reach the speed of light. The absoluteness of the speed of light means that *you* can never keep up with your own light, and because everyone else will agree that your light is beating you, they will always conclude that you are traveling slower than the speed of light.

allowing us to see those galaxies as they were in the distant past. Indeed, powerful telescopes now often observe galaxies for which this is the case. However, as long as the distance between us and them is growing at a rate greater than the speed of light, there is no possible way for anyone or anything to travel from one to the other.

Therefore, there is still no *measurement* that anyone can make in which you would be outracing light. Again, what relativity tells us isn't that "nothing can travel faster than light," but rather that nothing can ever accelerate to a point at which it would outrace light.

Those of you who are familiar with some of the strange effects of quantum mechanics may also know of another often claimed loophole—namely, that under certain circumstances known as "quantum entanglement," measuring a particle in one place can affect a particle in another place *instantaneously*. However, while laboratory experiments suggest that this instantaneous effect can really happen, current understanding of physics tells us that it cannot be used to transmit any useful information from one place to the other; indeed, if you were at the location of the first particle and wanted to confirm that the second had been affected, you'd need to receive a signal from its location, and that signal could not travel to you faster than light. To be clearer about such cases, physicists often prefer to say that the speed of light is a limit to the speed of transmission of information, but in my mind this is still the same as saying that nothing can outrace light.

THE SPRINTER AND THE LIGHT BEAM

Just to hammer home the surprising and amazing implications of light's absolute speed, I'll give you one more example. Imagine a future track champion; let's call him Ben. Shortly after breaking the world record in the 100-meter dash, Ben is accused of having used banned performance enhancements. Being an honest type, he admits his transgression, but his evident lack of remorse leads to his being banned from official competitions. He therefore decides to focus on continuing to "improve" himself through medicine, and on training harder than ever. One day, he holds a press conference to announce that since he's banned from competing against people, he intends to race a beam of light!

Ben's announcement generates enormous excitement, and he has no trouble lining up sponsors. Finally, race day arrives, and the starting gun goes off in a sold-out stadium. Ben races out of the starting blocks with superhuman speed, shattering the 100-meter world record by running the distance in 8 seconds flat. Unfortunately for Ben, the crowd is not impressed. The light beam, having emerged from its "blocks" at the speed of light, covered the same distance in less than a millionth of a second. Ben goes home, disgraced once more. But he is not the type to give up easily.

Over the next two years, he trains in secret, testing all manner of experimental performance enhancements. Nearly forgotten by the public, he finally reemerges, announcing, "I'm ready for a rematch." Sponsors are hard to come by this time, and spectators are scarce on race day. But for those few who are there to witness it, an incredible spectacle unfolds. As the starting gun goes off, Ben blasts out of the starting blocks at 99.99% of the speed of light and maintains this speed to the finish. Because this means the entire race is over in a millionth of a second, the excitement comes in watching the super-slow-motion replay.

The replay shows that the light, of course, goes start to finish at the full speed of light and therefore wins again—but barely! Because Ben's speed was only $0.0001c$ slower than the light's speed, the light gradually pulls ahead of Ben as both move down the track, and the light crosses the finish line only one centimeter ahead of Ben.

The crowd goes wild, and TV reporters rush to find Ben for an interview. But he seems to be missing. Finally, a reporter checks the locker room and finds Ben sulking under a towel. "What's wrong?" asks the reporter. Ben turns, and with tears in his eyes, he says, "Two years of training and experiments, and the light beat me just as badly this time as last time."

You probably realize what has happened. From the crowd's point of view, the race with light was very close, because Ben was going only very slightly slower than the speed of light. But from Ben's point of view, the absoluteness of the speed of light meant that he would have seen the light beam going the full speed of light faster than him. In other words, the light was just as much faster than him as it was two years earlier. The only consolation he'll get is that, to his surprise, he'll have found that as he ran, the race course was unexpectedly short. But that's a topic for the next chapter.

3

REDEFINING SPACE AND TIME

THE FORESHORTENED race course that Ben experiences while running comes about for the same reason that you found the distance to the black hole shorter during your imaginary journey in chapter 1. It is also closely related to the reason why less time passed for you than for people on Earth during your trip, as well as to other effects described by Einstein's special theory of relativity. These include the idea that observers in different reference frames may disagree about whether two events are simultaneous, and the famous equivalence of mass and energy embodied in the equation $E = mc^2$.

In this chapter, we'll use more thought experiments to show that all of these amazing ideas are direct consequences of the absoluteness of the speed of light. As we'll see, keeping the speed of light constant requires a fundamental change to our old conceptions of space and time. Let's begin by examining how the constancy of the speed of light affects time.

SPEED, DISTANCE, AND TIME

In our daily lives, we expect observers in different reference frames to get different results for the *speeds* of moving objects, but to agree about travel times. As a simple example, imagine that, with the aid of a tape measure, you carefully measure the distance you walk from home to work and find it to be 10.0 kilometers. You then get on your bicycle and find that it takes you 30 minutes, or ½ hour, to ride from home to work. Because speed is defined as distance divided by time, you would conclude that your speed on your bicycle is 10.0 kilometers ÷ ½ hour = 20 kilometers per hour. Now, consider how your trip would appear to someone watching from the Moon. As we discussed earlier (see figure 2.1), observers on the Moon would see you traveling much faster than 20 kilometers per hour, because they would also see you moving with the speed of Earth's rotation; for example, if you were at the equator and traveling west to east (the direction in which Earth rotates), they would say that you are moving at 1,690 kilometers per hour (your biking speed of 20 km/hr plus Earth's equatorial rotation speed of 1,670 km/hr). This much higher speed would make sense, because they also would see you move a longer distance; during your half-hour ride, they'd see you move a distance of half of the 1,690 kilometers that you would move in an hour, or 845 kilometers. With a little calculation to account for the effect of Earth's rotation, they'd also be able to conclude that the distance from your home to work along Earth's surface is 10.0 kilometers, in agreement with your measurement of the distance.

But what happens if we switch to *light* making the trip from your home to your work? For example, suppose you turn on a flashlight and time how long it takes the light to travel from home to work (which you might accomplish, for example, by having a mirror at work and measuring the round-trip time, then dividing by two). This time, the observers on the Moon must measure the light beam's speed to be exactly the same speed of light that you measure; that is, the fact that they observe you to be moving with Earth's rotation would not change the speed they measure for the light beam.

Now we come to the key point: The observers on the Moon would again see the light beam travel a greater distance than you see it travel, because of its motion along with the rotating Earth. However, because they measure the speed to be exactly the same speed that you measure—and because speed is always equal to distance divided by time—they would be forced to conclude that the time it takes for the light to travel from your home to your work is *greater* than the time that you say it takes the light to make the trip. (Just to be sure this is clear: Because speed = distance ÷ time, the only way to calculate the same speed with a larger distance is if you divide this larger distance by a larger time.) Moreover, because you and the observers on the Moon disagree on the time it takes light to travel from your home to your work but agree on the speed the light travels, you must also disagree on the distance from your home to your work as measured along Earth's surface. In other words, the fact that everyone always agrees on the speed of light means that we no longer all agree on measurements of time and distance. The disagreements would be very small in this case, because Earth's rotation is so slow compared to the speed of light.[1] The effect will become much more dramatic if we now switch to a high-speed thought experiment.

TIME DILATION

You and Al are both back in your spaceships, floating freely. You feel yourself to be at rest, and see Al moving past you at a high speed. Al, of course, says that he is the one at rest and you are moving past him at high speed. No problem so far, as your differing points of view simply reflect the fact that all motion is relative. Now, imagine that Al

---

1. In case you are curious about how much the difference would be: The time that you measure for the light to travel the 10.0 kilometers from home to work would be about 33 microseconds (which is 10 km divided by the speed of light). If you multiply this 33 microseconds by the speed of Earth's rotation, you'll find that Earth's rotation would carry you a distance of only about 15 millimeters during the light's trip from your home to your work. In other words, the observer on the Moon would see you travel only about an extra 15 millimeters due to Earth's rotation, which is so small compared to the 10-kilometer distance that you measure from home to work that it has only a very tiny effect on the way you and the Moon observer would measure the distance and the time.

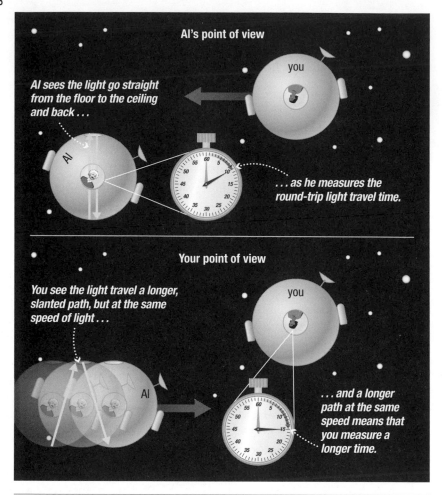

FIGURE 3.1 Al shines a laser from his floor to a ceiling mirror and back. His forward motion means you will see the light trace out a longer, triangular path. Because you and Al agree on the light's speed, you must measure it to take a longer time to complete that longer path.

has a laser on his floor that is pointed up to a mirror on his ceiling. As shown in the top panel of figure 3.1, he momentarily flashes the laser light so that the beam travels straight up to the ceiling mirror, which reflects it back down to the floor. Using very accurate clocks, each of you measures the round-trip time for the light to go from Al's floor to his ceiling and back. What will you find?

From your point of view, Al's high speed will move his entire ship noticeably forward during the time that the light travels from floor to ceiling and back. As a result, you'll see the laser beam trace out the triangular path shown in the bottom panel of figure 3.1. By our pre-relativity thinking, this would be no big deal. You and Al would agree on how long it takes the light to go from the floor to the ceiling and back, but you would disagree on the speed at which the light makes its round-trip journey. But this cannot be the case, because everyone always measures the same speed of light.

To understand what *really* happens, remember that at any given speed, traveling a longer distance must take a longer time. For example, if you travel at a speed of 100 kilometers per hour, it will obviously take you longer to travel 15 kilometers than it takes you to travel 10 kilometers. Returning to our spaceships, you and Al will agree that the light is traveling at the speed of light, or 300,000 kilometers per second. Therefore, because the triangular path that you see the light take is clearly longer than the straight up-and-down path that Al sees it take, *you must measure a longer time* for the light's round-trip. In other words, if you look at Al's clock while all this unfolds, you must see it running at a *slower rate* than yours, since that is the only way that his clock can end up showing less elapsed time than your clock during the light's round-trip.

Note that it doesn't matter what kind of "clock" you and Al use to measure the time for the light's round-trip. You'll get the same result whether you measure the time with mechanical clocks, electrical clocks, atomic clocks, heartbeats, or biochemical reactions. In every case, you'll observe Al's clocks to be running slower than yours. Our astonishing conclusion: From your point of view, *time itself* is running slower for Al.

How much slower is time running for Al? It depends on his speed relative to you. If he is moving slowly compared to the speed of light, you will scarcely be able to detect the slant of the light path, and your clock and Al's clock will tick at nearly the same rate. That is why we don't notice these effects in our daily lives, in which even spaceship speeds are only a minuscule fraction of the speed of light.

The slant of the light path—and hence the slowing of time— becomes noticeable only as Al's speed begins to approach the speed of light. The faster he travels, the more slanted the light path will appear to you, and the greater the difference between the rate of his clock and that of yours.

This effect, in which time runs slower in reference frames that are moving relative to you, is called *time dilation*; the term comes from the idea that time is dilated, or expanded, in a moving reference frame. The faster the other reference frame is moving, the more slowly you'll see time to be passing within it.

In case you'd like to be more precise, it's easy to calculate the factor by which time slows down in a moving reference frame. Just follow these three simple steps:[2]

1. Write the moving object's speed as a fraction of the speed of light.
2. Square the fraction and subtract it from 1.
3. Take the square root.

For example, suppose that Al is moving past you at 90% of the speed of light, or 0.9c. The first step tells us to use the fraction, 0.9. The second step tells us to square this fraction, which gives us $0.9^2 = 0.81$, and then subtract it from 1, which gives us $1 - 0.81 = 0.19$. The last step tells us to take the square root, and with a calculator you'll find that $\sqrt{0.19} \approx 0.44$. This tells us that, as Al moves by you at 0.9c, you'll observe time to be passing only 44% as fast for him as for you. In other words, if you could continue to observe Al while 10 seconds go by for you, you would observe his clocks passing only 4.4 seconds; similarly, while 100 years pass for you, you would observe that only 44 years pass for him.

---

2. If you don't mind equations, it's easier to put the three steps into the following single formula, which can be derived simply by applying the Pythagorean theorem to the laser light triangle shown in figure 3.1:

Time in the moving reference frame = (Time in your rest frame) $\times \sqrt{(1 - (v/c)^2}$

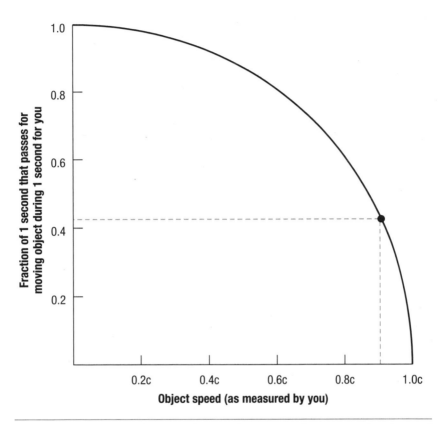

FIGURE 3.2 This graph shows the slowing of time as an object approaches the speed of light. The dashed lines show that, at 0.9*c*, time is slowed by a factor of 0.44. At higher speed, time is slowed even more. In principle, time would come to a halt for an object moving at the speed of light.

We can apply these ideas more generally by making a graph of how time slows down as an object's speed increases; the result is shown in figure 3.2. The horizontal axis shows the object's speed as measured by you, while the vertical axis shows how much time (in seconds) passes for the object during each of your seconds. At speeds that are low compared to the speed of light, the amount of time that passes for the moving object is almost indistinguishable from 1 second, which means that time for you and time for the moving object will pass at nearly the same rate. But as the object's speed increases, its time slows noticeably. For example, the dashed lines show the result we found above, which is that at $0.9c$, you will observe only 0.44 seconds passing for Al during each of your seconds. Notice that the graph drops toward zero as the speed approaches the speed of light, which means that time slows more and more as an object gets closer and closer to the speed of light.

The slowing of time offers another way of looking at the fact that no object can ever accelerate enough to reach the speed of light. Imagine watching a spaceship accelerating away from you, so that its speed continually increases as the pilot keeps firing the engines. As the ship approached the speed of light, its time would run ever slower, which means you'd see its engines firing ever more slowly and weakly, even if the pilot was always firing them at full power. Although the spaceship's speed could get closer and closer to the speed of light, it could never quite get that last engine boost needed to reach the speed of light, because time would in principle come to a complete stop at that speed. In a sense, the spaceship could never reach the speed of light because it just wouldn't have enough time.

LENGTH AND MASS

The fact that time is different in different reference frames automatically implies that distances (or lengths) and masses must also be affected by motion. Let's start with length, by going back to our sprinter, Ben, as he races the light beam.

Because Ben is traveling down the 100-meter track at a speed very close to the speed of light, his time must be running much slower than the time for us and other people watching him in the stadium. In fact, given that he is running at a speed of $0.9999c$, our calculation recipe from above tells us that time for him is passing only 1.4% as fast as time for the spectators (because $\sqrt{(1 - 0.9999^2}$ $= 0.014)$. Therefore, from his point of view, he can cover only 1.4% as much distance as we would say he's covering. For him, the 100-meter race is only 1.4 meters long while he runs.

We can use the same idea to explain what happened on your journey to the black hole in chapter 1. Recall that you made that trip at 99% of the speed of light, or 0.99c. The calculation recipe tells us that your time runs only about 14% as fast during your journey as it does for those of us who stay home on Earth. Because your journey has you traveling at this high speed for 50.5 years as we view it on Earth, only 14% of that, or about 7 years, passes for you. (The 6 months that you spend orbiting the black hole are essentially the same for both you and those of us on Earth, since you orbited far enough from the black hole that its gravity would have only a very small effect on your time.) Moreover, you will agree with us about your relative speed of travel; just as we say that you are moving away from Earth at 0.99c on your outward journey, you'll say that Earth is moving at 0.99c away from you. Therefore, since you will make the journey in only 14% as much time as we say it takes, you will also measure the distance you travel to be only 14% as long. That is why the 25 light-years to the black hole shrank to only about 3½ light-years when you got up to speed.

The idea that the traveler will measure a shorter distance can be turned around to lead to a closely related conclusion. Remember that motion is always relative, and all viewpoints are equally valid. For example, from Ben's point of view on the race course, he's just moving his legs while the race course goes by under him. Therefore, just as he finds the course shrunk in the direction in which it goes by, we will find that *he* is shrunk in the direction in which we see him go by.[3]

---

3. It's important to realize that while careful study of the situation will lead us to conclude that Ben is shrunk in his direction of motion, we would not actually *see* him to be "flattened out" when we looked at him. The reason is that at such high speeds, what we see is also affected by differences in light travel time from different parts of his body to us at different points along the race course. Numerous websites offer simulations of what objects would actually look like when moving by us at high speed.

In other words, we would measure Ben to be flattened in the direction that he's moving. (His height and width will not be affected.) In the same way, if the spaceship that you take to the black hole is 100 meters long when it is at rest, we'll find it to be shorter than that while you are traveling at high speed. For this reason, the effects on distance and length in relativity are usually called *length contraction*. The factor by which length contracts for a moving object is the same as the factor by which we see its time dilated, so you can use the graph in figure 3.2 for length as well as for time.

Turning to mass, relativity tells us that a moving object behaves as though it has a *greater* mass than the same object at rest.[4] If you want to know how much greater, just *divide* by the factor that we found for time dilation and length contraction. For example, we've found that during your journey to the black hole at 0.99c, we'll see your time running only 14% as fast as ours, and measure your spaceship to be only 14% as long as it is at rest. Therefore, because 14% is the same as 0.14, we would say that your mass seems to be $1 \div 0.14 \approx 7.1$ times as large as it is at rest. In other words, if you have a normal rest mass of 50 kilograms, then your mass would seem to increase to about $50 \times 7.1 = 355$ kilograms when you are moving at 99% of the speed of light. As always, remember that *you* will still think you have your normal mass, because you can consider yourself to be the one at rest. It is only those of us observing you who would find your mass to be higher; for example, if you crashed into something, the force of the crash would be about 7 times greater than we would expect if we assumed your

---

4. The concept of mass increase is in some sense an oversimplification, because what we actually observe for a moving object is its momentum and energy, and in the mathematical treatment of relativity these become combined into what is often called "momenergy" (or "four-momentum"). For this reason, although "mass increase" was taught as a part of relativity for many decades, most physicists today prefer to think in terms of the increase affecting momenergy, which allows them to treat mass as an invariant quantity that does *not* increase with speed. This distinction will be important if you study relativity further, but for the "intuitive introduction" in this book, I believe it is still easier to use the older approach of thinking in terms of mass increase.

normal mass, thereby confirming that your mass seems to be greater by a factor of about 7 while you are moving.

Why does mass appear to increase in this way? There are several equivalent ways to look at it, but I find it clearest to use another thought experiment. Imagine that Al has an identical twin brother with an identical spaceship, but the brother is at rest in your reference frame while Al is moving by you at high speed. At the instant Al passes by, you give both Al and his brother identical pushes; that is, you push them both with the same force for the same amount of time. By our pre-relativity thinking, in which both spaceships would have the same mass, you would expect your push to accelerate both Al and his brother by the same amount, such as gaining 1 kilometer per second of speed relative to you. But now think about what is happening to time: Because Al is moving relative to you and his brother, Al's time will be running more slowly than yours and his brother's—which means that he experiences your push for a shorter time than does his brother. Because he experiences the push for a shorter time, it will have a smaller effect on him, accelerating him less than it accelerates his brother. One way to explain how your identical pushes can cause a smaller acceleration for Al is if his mass is greater than his brother's mass.[5]

The idea of mass increase provides still another way of explaining why no material object can reach the speed of light. The faster an object is moving relative to you, the greater the mass you'll find it to have. Therefore, as an object's speed gets higher and higher, the same force will accelerate the object by smaller and smaller amounts. As the object's speed approaches the speed of light, you would find its mass to be heading toward infinity. No force can accelerate an infinite mass, so the object can never gain that last little bit of speed needed to push it to the speed of light.

---

5. Again, keep in mind that the concept of "mass increase" is no longer commonly used by physicists who study relativity, but this distinction will be important only if you study relativity further; it will not affect our "intuitive introduction."

THE RELATIVITY OF SIMULTANEITY

We've covered the most famous consequences of special relativity, which are the ideas that objects in moving reference frames have (1) slower time, (2) contracted length, and (3) increased mass. From these three ideas, it is possible to come up with many other amazing or seemingly paradoxical consequences of special relativity. We can't possibly cover them all in a short book like this one, but I'd like to call your attention to one that will be especially important when, in the next chapter, we try to redefine our "common sense" to accommodate relativity. It is the idea sometimes known as the *relativity of simultaneity*.

Our old common sense tells us that everyone must agree on whether two events happen at the same time or one event happens before another. For example, if you see two apples—one red and one green—fall from two different trees and hit the ground at the same time, you expect everyone else also to agree that they landed at the same time (assuming that you've accounted for any difference in the light travel times from the two trees). Similarly, if you saw the green apple land before the red apple, you would be very surprised if someone else said that the red apple hit the ground first. Well, prepare yourself to be surprised.

Before I surprise you, however, it's important to be clear on what can be relative and what cannot. Although we'll find that observers in different reference frames do not necessarily agree about the order or simultaneity of events that occur in different places, everyone must still agree about the order of events that occur *in a single place*. For example, if you grab a cookie and eat it, everyone must agree that you ate the cookie only after you picked it up.

Now to our thought experiment. Al has a brand-new, extra-long spaceship, and he is coming toward you at a speed of 90% of the speed of light, or $0.9c$. He is in the center of his spaceship, which is totally dark except for a flashing green light at its front end and a flashing red light at its back end. Suppose that, as shown in the top panel of figure 3.3, *you* see the green and red lights flash at exactly the

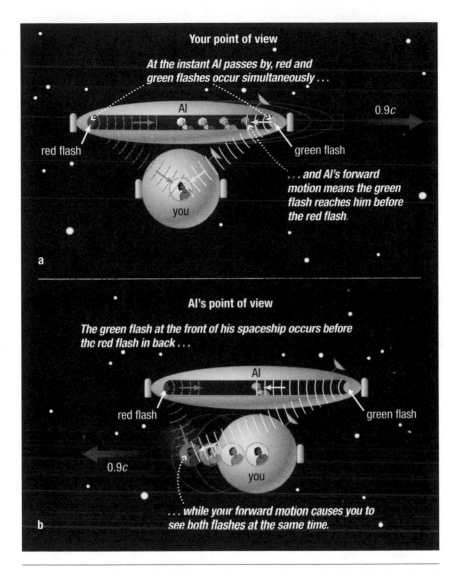

FIGURE 3.3 The green and red lights reach you simultaneously while Al is illuminated first in green and then in red. You will conclude that both lights actually flashed simultaneously, but Al will conclude that the green light flashed first.

same time, with the flashes occurring at the instant that Al happens to pass you, so that the two lights both illuminate you at the same instant. Notice that, during the very short time that the light flashes are traveling toward you, Al's forward motion will carry him *toward* the point where you saw the green flash occur. As a result, the green flash will reach him before the red flash, which means he'll be illuminated first by green light and then by red light. So far, nothing should be surprising: You see the two flashes simultaneously because you are stationary, but the green flash reaches Al before the red flash because of his forward motion. But now let's think about what happens from Al's point of view, in which he is the one who is stationary and you are the one who is moving.

Remember that motion cannot affect the order of events that occur in the same place. Therefore, since you are in a single place at the moment when both the green light and the red light reach and illuminate you, everyone will agree that both lights reached you at the same time; it's the same idea as everyone agreeing that you pick up the cookie before you eat it. Similarly, everyone, including Al, will agree that he is illuminated first by the green light and then by the red light. However, Al considers himself to be stationary in the center of his spaceship, with the green and red lights located equal distances away. Therefore, from his point of view, the only way that the green light can reach him before the red light is if *the green flash actually occurred first*. In other words, he'll see the situation as shown in the bottom panel of figure 3.3: He'll say that the green flash occurred before the red flash, and that the reason they both reached you at the same time was because you were moving in the direction of the red flash.

Other perspectives are also possible. For example, a person in a ship moving in the opposite direction (in your reference frame) would conclude that the red flash occurred first. We therefore see at least three different viewpoints on the order of the events. You say both flashes were simultaneous, Al says the green occurred before the red, and an observer going the other direction would say the red occurred before the green. Who is right? All motion is relative, so you are all equally correct. Our new common sense will need to allow for the fact that not all observers will agree on the order or simultaneity of events.

SPACETIME

All this talk about disagreement on time, length, mass, and even the order of events may be starting to make you wonder if "everything" isn't relative after all. But there are some patterns that you may see emerging. We can think of length as a measurement of space (since space has length, width, and depth), and while time and space may differ for different observers, they do so in very precise ways, ensuring that everything remains self-consistent from any observer's point of view. If you think about it, this self-consistency is really just the first absolute of relativity, that the laws of nature are the same for everyone. And remember that we've found all of these results by focusing on the second absolute, that everyone always measures the same speed of light.

Einstein and others investigated how these ideas manifest themselves mathematically, and they discovered something very important: While measurements of space and time may independently differ for different observers, the combination of space and time known as *spacetime* is the same for everyone.

To understand what we mean by spacetime, you must first understand the concept of *dimension*. We can define dimension as the number of independent directions in which movement is possible. A point has zero dimensions, because a geometric prisoner confined to a point would have no place to go. Sweeping a point back and forth along one direction generates a line. The line is one-dimensional because only one direction of motion is possible (going backward is considered the same as going forward by a negative distance). Sweeping a line back and forth generates a two-dimensional plane. The two directions of possible motion are, say, lengthwise and widthwise.

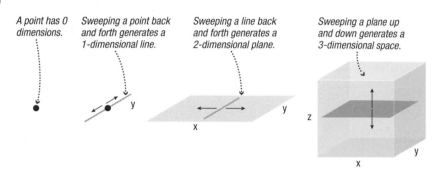

A point has 0 dimensions.

Sweeping a point back and forth generates a 1-dimensional line.

Sweeping a line back and forth generates a 2-dimensional plane.

Sweeping a plane up and down generates a 3-dimensional space.

FIGURE 3.4 These diagrams show how we build up three-dimensional space.

Any other direction is just a combination of these two. If we sweep a plane up and down, it fills three-dimensional space, with the three independent directions of length, width, and depth. Figure 3.4 summarizes these ideas.

We live in three-dimensional space and therefore cannot visualize any direction that is distinct from length, width, and depth (or combinations thereof). However, just because we cannot see "other" directions doesn't mean they don't exist. If we could sweep space back and forth in some "other" direction, we would generate a four-dimensional space. We cannot visualize a four-dimensional space, but it's relatively easy to describe it mathematically. In algebra, we do one-dimensional problems with the single variable x, two-dimensional problems with the variables x and y, and three-dimensional problems with the variables x, y, and z. A four-dimensional problem simply requires adding a fourth variable, as in x, y, z, and w. In principle, we can continue to five dimensions, six dimensions, and so on.

Any space with more than three dimensions is called a *hyperspace*, which simply means "beyond space." Spacetime is a particular hyperspace in which the four directions of possible motion are length, width, depth, and *time*. Note that time is not "the" fourth dimension; it is simply one of the four. Although we cannot picture all four dimensions of spacetime at once, we can imagine what things would look like if we could. In addition to the three spatial dimensions of spacetime that we ordinarily see, every object would be stretched out through time. Objects that we see as three-dimensional in our ordinary lives would appear as four-dimensional objects in spacetime. If we could see in four dimensions, we could look through time just as easily as we look to our left or right. If we looked at a person, we could see every event in that person's life. If we wondered what really happened during some historical event, we'd simply look to find the answer.

The concept of spacetime provides a simple way to understand why different observers can disagree about measurements of time and distance. Because we can't visualize four dimensions at once, we'll use a three-dimensional analogy. Suppose you give the same book to many different people and ask each person to measure the book's dimensions. Everyone will get the same results, agreeing on the three-dimensional structure of the book. Now, suppose instead that you show each person only a two-dimensional picture of the book rather than the book itself. The pictures may look very different, even though they all show the same book (figure 3.5). If the people believed that the two-dimensional pictures reflected reality, they might each measure the book's length and width differently, drawing different conclusions about what the book really looks like. In our ordinary lives, we perceive only three dimensions, and we assume that this perception reflects reality. But spacetime is actually four-dimensional. Just as different people can see different two-dimensional pictures of the same three-dimensional book, different observers can see different three-dimensional "pictures" of the same spacetime reality. These different "pictures" are the differing perceptions of time and space of observers in different reference frames. That is why different observers can get different results when they measure time, length, or mass, even though they are all actually looking at the same spacetime reality. To quote the words of E. F. Taylor and J. A. Wheeler in their classic textbook, *Spacetime Physics*:

Space is different for different observers.
Time is different for different observers.
Spacetime is the same for everyone.

A book has an unambiguous three-dimensional shape.

Two-dimensional pictures of the book can look very different.

FIGURE 3.5 Just as a three-dimensional object can seem different depending on the two-dimensional picture taken of it, the reality of an object in spacetime may lead different observers to different measurements when they look at space and time independently.

I'll add just one more note about spacetime for now, directed mainly at readers who may be trying to visualize four-dimensional pathways and who don't mind a bit of mathematics. (Others can skip this paragraph.) Suppose that you draw a point at the origin (center) of a graph, then draw a second point some distance away. Depending on the directions in which you draw your x (horizontal) and y (vertical) axes, you can get different answers for the x and y coordinates of the second point. However, no matter what you do, you'll always get the same answer for the *distance* between the two points, which will be $\sqrt{(x^2 + y^2)}$. In the same way, the distance between the origin and any point in three-dimensional space is always $\sqrt{(x^2 + y^2 + z^2)}$, no matter how you orient the three axes. Because spacetime is the same for everyone, there must also be a spacetime "distance" between two events, more formally called the *interval*, that everyone always agrees on regardless of how they measure space and time individually. You might expect the interval to look just like our earlier equations for distance but with an added t^2 under the square root. However, the interval turns out to look slightly different; it is $\sqrt{(x^2 + y^2 + z^2 - t^2)}$.[6] The minus sign introduces added complexity to the geometry of spacetime. For example, while the three-dimensional distance between two points can be zero only if the points are in the same place, the interval between two events can be zero even when they are separated in spacetime, as long as they can be connected by a path that represents the four-dimensional path of a light beam. We won't go into the details of this geometry in this book, but you'll encounter it if you study relativity further.

THE FAMOUS EQUATION $E = MC^2$

We can extend our book analogy (figure 3.5) to gain a little deeper insight into length contraction, time dilation, and mass increase. Let's try to picture two books in spacetime, one that is stationary in your reference frame and the other that is moving by you at high speed. To keep things simple, let's picture the books over a period of just one hour of your time, and assume that both were right in front of you at the start of the hour.

---

6. Mathematically inclined readers will notice that t has time units while x, y, and z have length units. To make the units consistent, you can replace t with ct; in this book, we'll assume that this is done implicitly.

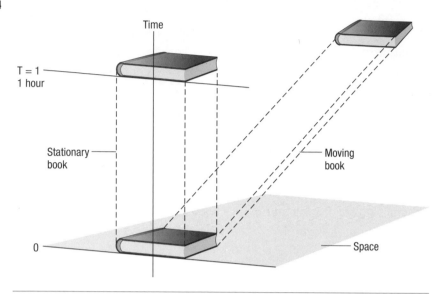

FIGURE 3.6 If we visualize a book through one hour of spacetime, a stationary book stays fixed in space but moves in time, while a moving book moves in both space and time.

We obviously can't picture all four dimensions of spacetime, so instead we'll represent space with a flat plane and then extend a time axis upward, as shown in figure 3.6. (Technically, this means we could show only two of the book's spatial dimensions, but we'll ignore this detail.) The book that is stationary in your reference frame simply extends straight up your time axis for a "distance" of one hour. The book that is moving by you at high speed starts from the same place, but it extends away from you in space during your hour of time.

Now, remember that although everyone must agree on each book's four-dimensional spacetime structure, we observe only three dimensions at once. When you observe the stationary book, you are moving up the time axis parallel to the book. You therefore see—or measure—only its three spatial dimensions, not its time dimension; in our previous analogy, it's rather like looking face-on at the book's cover, so that you see the length and width of the cover but not the thickness of the book. In contrast, observing the moving book is rather like observing it when it is rotated at some angle. Just as a two-dimensional photo of the angled book will make the cover look smaller in the direction of the rotation, your three-dimensional observations of the moving book will make one of its three spatial dimensions smaller—which is the phenomenon of length contraction. Moreover, just as the rotation means a photograph will now show some of the book's thickness that you did not see before, the spacetime "rotation" means that you can now observe part of the moving book's time dimension, an idea that manifests itself as time dilation.

So far, so good, but what about mass increase? In pre-relativity physics, there were two quantities that were always thought to be conserved independently: mass and energy. That is, scientists assumed that if you had a closed system (one unaffected by outside forces), its total mass would always remain the same and its total energy would always remain the same. But relativity shows that mass changes with motion, which means that it cannot be conserved by itself; instead, it must be some combination of mass and energy that is conserved. Let's look at this idea in spacetime.

You are probably familiar with the idea that moving objects have what we call kinetic energy; the faster an object moves, the greater its kinetic energy.[7] This means that in pre-relativity physics, a non-moving object has no energy. In relativity, however, we cannot ignore time, and all objects are essentially always moving through time. Moreover, because time is not "the" fourth dimension but just one of the four spacetime dimensions, there's no reason to think that time should be ignored when we consider an object's energy.

Einstein worked through this idea (albeit in a somewhat different way) with his equations of special relativity and discovered that there is indeed an extra component to energy, beyond the normal kinetic energy, that had not previously been recognized. He found that for a moving object, this extra energy manifests itself as mass increase,

---

7. The nonrelativistic formula for an object's kinetic energy is $\frac{1}{2} mv^2$, where m is the object's mass and v is its velocity or speed.

which can be expressed with a simple formula. Perhaps more surprisingly, he also found that it implies that there is an energy associated with "moving through time" even for objects that are not moving in space—that is, for objects at rest. Although we won't go through it in this book, relatively simple algebra allows you to calculate this rest energy from the mass increase formula, and voilà! it turns out to be mc^2, where m is the object's rest mass (its mass measured in a reference frame in which it is stationary) and c is the speed of light. When we include an E to stand for energy, we get what is probably the most famous equation in the world: $E = mc^2$.

This equation tells us that it is possible, at least under certain circumstances, to convert mass into energy and vice versa. Moreover, the factor c^2 represents a very large number (in standard units, $c^2 = 300{,}000{,}000^2 \text{ m}^2/\text{s}^2 = 9 \times 10^{16} \text{ m}^2/\text{s}^2$), which tells us that a small amount of mass can yield an enormous amount of energy. For example, the energy released by the atomic bombs used in World War II came from converting barely one gram of mass—about as much mass as a paper clip—into energy. $E = mc^2$ also explains how the Sun shines, which is by continuously converting a tiny fraction of its mass into energy through the process of nuclear fusion.

More generally, $E = mc^2$ expresses a type of equivalence between mass and the energy it contains when at rest. Keep in mind that you must view this equivalence in much the same way that we view the equivalence of space and time. That is, while we may know that space and time are just different dimensions of a single spacetime reality, space and time *look* very different to us in our everyday lives. Similarly, mass and energy look different to us under most circumstances, and we notice an equivalence only rarely in our daily lives. Nevertheless, those rare circumstances in which we can observe the equivalence, such as with nuclear bombs or the shining of stars, prove beyond doubt that Einstein's famous equation has profound effects on our existence. They also show that special relativity, from whence this equation came, is a deeply important theory that affects each and every one of us every day.

4

A NEW COMMON SENSE

IN CHAPTERS 2 and 3, we saw how all the major consequences of Einstein's special theory of relativity flow from two simple absolutes: that the laws of nature are the same for everyone and that everyone always measures the same speed of light. We found that you can't outrace your own light. We found that observers in different reference frames, by which we mean frames that are moving relative to each other, will come up with differing measurements for time, space, and mass. We found that different observers won't necessarily agree on the order or simultaneity of two events that occur in different places. And we saw that an object's rest mass and energy have a type of equivalence described by Einstein's famous equation $E = mc^2$.

I hope you'll agree that none of this has been particularly difficult. We've done nearly all of it with just a few thought experiments, and we've used very little mathematics. Yet, as easy as this has been, you're probably still thinking, "Huh?" After all, it's one thing to work through the logic that leads to the astonishing consequences of relativity, and another to claim that they "make sense." So in this chapter I'll attempt to help you take the ideas you've already learned and make some sense of them.

Before we begin, it's worth noting that, contrary to a common belief, special relativity does not really violate common sense. The differences between what we expect in everyday life and what relativity tells us become obvious only when we deal with objects moving at speeds close to the speed of light, and such speeds are not part of our common, everyday experiences. We can't possibly have "common" sense about things that we don't commonly experience.

The real problem in making sense of relativity is that we tend to assume that our low-speed common sense should also apply at high speeds. But why should we think that? After all, there are many other cases in which something that we learn for a limited set of circumstances turns out to require modification to apply in broader circumstances.

Consider the meaning of "up" and "down." At a very young age, you learned "common sense" meanings for up and down: Up is above your head, down is toward your feet, and things tend to fall down. This common sense worked perfectly well for you as a young child, and it still works fine when you apply it in your home or your community. One day, however, you learned that Earth is round, and you saw how we can represent Earth with a globe. You may not recall it, but this probably created a mini-intellectual crisis for you (figure 4.1).

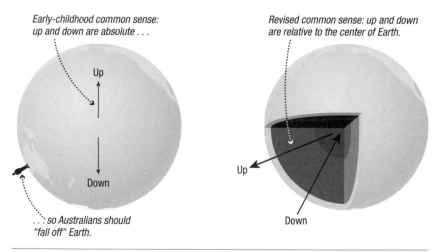

FIGURE 4.1 Your early "common sense" about up and down, which was based only on experiences in small parts of the globe, had to be revised to accommodate the fact that Australians don't fall off. In the same way, your low-speed common sense about space and time must be revised to accommodate what relativity tells us about higher speeds.

After all, if the globe had the Northern Hemisphere on top, then your common sense told you clearly that Australians would have to be falling off Earth. Since you knew that Australians don't fall off, you were forced to accept that your "common sense" about up and down was incorrect. You therefore revised your common sense, so that you recognized that up and down are really determined relative to the center of Earth, and seem like absolutes in space only when you look at just small pieces of Earth's surface.

Making sense of relativity requires the same type of revision to your common sense. Your low-speed common sense about space and time is fine as far as it goes, much as it's fine to think of up and down as absolute concepts in a game of basketball. But just as up and down must be redefined when you look at Earth as a whole, time and space must be redefined if you want to look at the full range of possible motions. It will take some mental effort, but it's really not that big a deal. After all, your new common sense will build upon your old one, and will still be perfectly consistent with everything that you commonly experience in your daily life.

THE RELATIVITY OF MOTION

In order to get you started in building a new common sense, I'm afraid that I first need to show you that some consequences of relativity may seem even more bizarre than they have so far. So get back in your spaceship for another high-speed thought experiment with your friend Al.

As you look out the window, you watch Al race away from you at a speed close to the speed of light. From our earlier thought experiments, we know that you'll say that his time is running slow, his length is contracted, and his mass is increased. But this time we need to ask a question that we've avoided up until now: What would Al say is going on?

As you know, Al would assume that he's the one at rest, and that *you* are moving away from him at high speed. Therefore, because the laws of nature are the same for everyone, he must reach *exactly the same conclusions* from his point of view that you reach from your point of view. That is, he'll say that *your* time is running slow, *your* length is contracted, and *your* mass has increased!

If you are like most of the students I've ever taught, you probably are not taking this well. After all, it seems contradictory; how can you both claim each other's time to be running slow? So like thousands of other space travelers before you, you decide to prove once and for all that it is Al's time that's running slow, not yours. This is easy to do: Just pull out a super telescope and observe what's going on in Al's spaceship. You'll find that his time is in fact running slow, and that everything he does unfolds in slow motion.[1] Armed with this visual evidence, you send a radio message to Al announcing your discovery: "Hey, Al! I'm watching, and your time is definitely running slower than mine."

Because the radio message travels at the absolute speed of light, Al has no trouble receiving your message (though its frequency will be reduced by the Doppler effect, since he is moving away from you). It takes the message a bit of time to reach him, of course, and a bit more time for his reply to reach you. And when it does, you hear him respond with a very slow-motion voice, "Hheeeelllllloooo tthhheeerrr . . . ," further verifying that his time is running slowly. But when you've finally recorded the entire transmission from Al, you play it back at higher speed so that it sounds normal, and to your utter surprise you hear him say: "What are you talking about? I'm watching you with my super telescope, and *you* are the one in slow motion!"

You can keep sending radio messages to argue back and forth all you want, but it will get you nowhere. Then you come up with a brilliant idea. You hook up a video camera to your telescope and record a movie showing that Al's clock is moving more slowly than yours.

---

1. Once again, I am avoiding the fact that what you *see* can be very different from what you would conclude after accounting for effects such as differences in light travel time (see footnote 3 in chapter 3). In this case, what you see is close to what you will conclude so long as Al is moving away from you, which is why I chose this setup for the thought experiment; you would *see* something quite different if Al were moving toward you or passing you by.

You burn your movie to a disc, put it into a very fast rocket, and shoot it off toward Al. You figure that when the movie arrives and he watches it, he'll have visual proof that you are right and that he really is moving in slow motion. Unfortunately, just before you can declare victory in the argument, you learn that Al had the same brilliant idea: A rocket arrives with a movie that he made. As you watch it, you see what appears to be clear proof that Al is right—his movie shows that *you* are in slow motion!

There is no way around it. The fact that the laws of nature are the same for everyone demands that as long you both are floating weightlessly in your spaceships, so that each of you can legitimately claim to be at rest, you must both conclude that the same things are happening to the other. You conclude that his time is running slow, he concludes that your time running slow, and that's the way it has to be.

A TICKET TO THE STARS

"Aha!" you say. "I've caught you this time! If Al and I stop moving and come together, we'll be able to compare clocks, and we can't both have recorded less time. Moreover, you told us earlier that *I* would be the one who aged less than the people on Earth on my trip to the black hole. So which one are you going to try to make me believe—that we would both say each other's clock is running slow, or that one of us really records less time?"

If I've guessed correctly about what you are thinking, you've discovered the famous *twin paradox* of relativity. In its standard form, you have an identical twin who stays home on Earth while you take a near-lightspeed journey to a star and back. Special relativity tells us that in both directions of the journey, your twin would conclude (after accounting for light-travel effects on what she actually sees) that your time runs slow on your spaceship, while you'd conclude that time runs slow back on Earth. This seems impossible, since you can't both be younger than the other when the two of you reunite at the end of the trip.

It does indeed present a paradox, but so did the fact that you thought people in Australia would "fall off" when you first learned about the round Earth. In other words, the paradoxes that we encounter in relativity are things that *seem* contradictory, but only when we apply our old common sense; they'll seem fine once we develop a new common sense. We'll get to that, but let's first resolve the paradox.

The key is to think about what it really means to say that the laws of nature are the same for everyone. If you are in an airplane on a smooth flight, you'll get exactly the same results for experiments that you'll get on the ground. But if you perform the experiments while the plane is taking off or flying through turbulence, you'll obviously get different results because of the forces acting on you and the experiments. You're still experiencing the same laws of nature, but in this case there are some forces acting on you that you didn't have to worry about when you were on the smooth flight. Therefore, if you want to compare your experiments to experiments done on the ground, you'll either have to take the additional forces into account or do the ground experiments in a flight simulator that creates the same forces. The practical consequence of this fact is that although the laws of nature are always the same for everyone, two observers will get the same results only when they are in equivalent reference frames; in other cases, the results will be more complex. That is why we've been dealing with free-float frames, in which both you and Al float weightlessly in your spaceships and therefore can each say that your circumstances are equivalent.

Let's think about how this comes into play on your trip to the black hole. We said that you made your trip to the black hole at a constant speed of $0.99c$ in both directions. As long as you are traveling at that constant speed, your reference frame is essentially equivalent to Earth's (neglecting Earth's relatively weak gravity), and it's therefore true that you would conclude that clocks are running slow on Earth. But we've neglected several critical issues: How did your spaceship go from being stationary on Earth to the speed of $0.99c$, how did it slow down to orbit the black hole, how did it accelerate again for the trip home, and how did it finally come to a stop back at Earth? We didn't say specifically (other than to note that the forces would kill you), but it's very clear that you were *not* in a reference frame equivalent to being at home during those periods of extreme acceleration (or deceleration). In other words, the rules we've discussed so far aren't enough to explain what happens when you feel the forces of acceleration.

We'll talk later about how general relativity can help us make sense of the twin paradox, but we can see its resolution with what we already know about spacetime. Your launch from Earth in the year 2040 is an event in spacetime, as is your arrival at the black hole and your return to Earth in the year 2091. Everyone must agree on the reality of these events, and the only question is how much time and space separates them. For someone on Earth, 51 years passed and your round-trip distance was 50 light-years. For you, traveling at 0.99c to and from the black hole, you found that only about 7½ years passed and that the round-trip distance was only about 7 light-years. This is nothing more than what we've already learned: Space is different for different observers, time is different for different observers, but spacetime is the same for everyone.

Now let's look at it in terms of your thought experiment with Al. It may be fine for you and him to argue back and forth by radio or rocketed videos about whose time is really going slow. But what happens if you get together and compare clocks? The answer depends on how you come together. To compare clocks, you need a starting point and an ending point. For the starting point, let's choose the instant when you and Al pass by each other, since you could in principle each hand off a clock to the other at that point. You can probably now see the problem: The ending point requires that you get together again, but as long as you continue in your equivalent, free-float reference frames, you'll each just see the other flying off into the universe at extremely high speed, never to return. The only way that you can get together to compare your clocks again is if one of you fires your rocket engines to (from the point of view of the other) slow down and come back. The technical details get a bit complex, but in the end it turns out that the one who fires the engines will experience forces much like those on your trip to the black hole, and will be the one with the clock that shows the shorter elapsed time.

In a sense, then, relativity offers a ticket to the stars, at least if we can ever really build spaceships capable of reaching speeds close to the speed of light. You've already used this ticket for your trip to the black hole. Without time dilation, you would have had to spend much of your life on a 51-year journey, but thanks to time dilation, you were able to make the trip in only 7½ years. If you made the trip at higher speed, time dilation would allow you to make it in even less time. For example, adding another 9 to make your speed 0.999c (rather than 0.99c) would reduce your round-trip time to the black hole to only about a year in each direction. In that case, you could leave Earth in 2040, and return home only two years older—but still in the year 2091, since people on Earth still would conclude that the trip takes you a little over 25 years in each direction, plus the time you spend at the black hole.

If we had technology to get you even closer to the speed of light, almost any trip could be made within your lifetime. For example, the Andromeda Galaxy is about 2.5 million light-years away, which means the round-trip to any star in the Andromeda Galaxy would take at least 5 million years from the point of view of people on Earth. However, if you could travel at a speed within 50 parts in 1 trillion of the speed of light (that is, at 0.99999999995c), the trip would take only about 50 years from your point of view. You could leave Earth at age 30 and return at age 80—but you would return to an Earth on which your friends, your family, and everything you knew had been gone for 5 million years.

So it's good news, bad news. The good news is that relativity offers a ticket to the stars, the bad news is that, in terms of time, it's a one-way ticket. You can go a long distance and return to the *place* that you left, but you cannot return to the time that you left. Relativity opens the universe to those who wish to travel, but there's no going home.

EXPERIMENTAL EVIDENCE FOR RELATIVITY

If you still think all this just sounds too weird, don't worry—so does everyone else when they first study relativity. It takes some getting used to, just like it took you a while to get used to your new ideas of up and down. If you've been able to follow the logic of the thought experiments, then you are doing as well as can be expected at this point.

Of course, before you accept the logic, you'd probably like to be sure that all of this is really backed by evidence. As we discussed earlier, all the logic in the world is not good enough to constitute evidence in science; we need actual observations or experiments. We've already discussed the evidence for the absoluteness of the speed of light. But how can we test other predictions of relativity?

The consequences of relativity are most noticeable at high speeds, so we'd like to conduct our tests with objects moving relative to us at speeds close to the speed of light. You might think that would be difficult, since *we* can't yet travel anywhere near those speeds. However, the objects need not be large, and it is relatively easy to get subatomic particles to reach such speeds. Scientists do this with the machines known as *particle accelerators*; these days, the Large Hadron Collider in Europe is the best-known accelerator, but scientists have been building similar machines (of lesser power) for many decades.

Particle accelerators may be complex and expensive machines, but their basic purpose is very simple: Scientists use them to accelerate subatomic particles to speeds near the speed of light and then crash the particles into each other, with the goal of observing the effects of the collisions. This simple purpose means that accelerators offer several direct tests of relativity.

First, the machines provide direct evidence that nothing can be accelerated to the speed of light. It is fairly easy to get particles traveling at 99% of the speed of light in particle accelerators. However, no matter how much more energy is put into the accelerators, the particle speeds get only fractionally closer to the speed of light. Some particles have been accelerated to speeds within 0.00001% of the speed of light, but none have ever reached the speed of light.

Second, accelerators allow us to test the prediction that mass should seem to increase. If we think in terms of pre-relativity physics, the amount of energy released when any two particles collide depends on the masses and speeds of the particles. We know the speeds of the colliding particles in an accelerator, so by measuring the energy of the collisions we can calculate the particle masses. The results show that the particles do indeed appear to have larger masses than they do when at rest, by exactly the amounts predicted by special relativity.

Third, accelerators offer a direct test of $E = mc^2$. Although this formula is most famous for showing how mass can be turned into energy (as in nuclear bombs), it also tells us that energy can be turned into mass. This is exactly what particle accelerators do. The collisions release highly concentrated energy, and some of this energy spontaneously turns into new subatomic particles. In fact, this is the main reason why scientists seek more powerful accelerators: With more energy, they can produce a greater array of new particles that may provide new insights into the building blocks of nature. From the standpoint of testing relativity, the mere fact that the particles are produced from energy verifies the predicted equivalence of mass and energy.

Fourth, and perhaps most remarkably, particle accelerators can provide a direct test of time dilation. Many of the particles produced from the energy of the collisions have very short lifetimes (or, more technically, short half-lives), meaning that they quickly decay (change) into other particles. For example, a particle called the π^+ ("pi plus") meson has a lifetime of about 18 nanoseconds (billionths of a second) when produced at rest. But π^+ mesons produced at speeds close to the speed of light in particle accelerators last much longer than 18 nanoseconds—and the amount longer is the amount predicted by the time dilation formula. Time really does run slow for these particles when they are moving relative to us at high speeds.

Other types of experiments have demonstrated the effects of relativity at lower speeds. Although effects such as time dilation become easy to notice only at very high speeds, in principle they are always present to at least some extent, and therefore can be measured with sufficiently accurate clocks. Over the past half century scientists have used the best available clocks to test relativity at increasingly low speeds. Time dilation has been measured by comparing clocks in the Space Shuttle and even in airplanes to clocks on the ground. In 2010, tests conducted at the National Institute of Standards and Technology in my hometown of Boulder, Colorado, verified the predicted amount of time dilation at speeds of less than 10 meters per second (36 kilometers per hour), which is slower than many of the bike riders cruising around town.

The bottom line is that special relativity is one of the most well-tested theories in all of science, and it has passed every test with flying colors. In science, we can never really prove a theory true beyond all doubt, since there is always the possibility that the theory will fail in some future experiment. Nevertheless, the huge body of evidence supporting the special theory of relativity cannot be made to go away, and if any other theory is ever to replace special relativity, it will still have to account for this body of evidence that supports the current theory so well.

SUNSHINE AND RADIOS

The direct experimental evidence for special relativity is impressive, but to my mind it isn't even the most important part. In particular, there are two somewhat indirect tests of relativity that play a vital role in our lives.

The first is the mass to energy aspect of $E = mc^2$. Besides explaining how nuclear bombs release so much energy, this equation also explains the energy produced in nuclear power plants, which provide a significant fraction (around 10% to 15%) of the world's electricity. Moreover, the conversion of mass to energy explains how the Sun and stars are able to shine steadily for millions to billions of years. In our Sun, for example, nuclear fusion transforms about 600 million tons of hydrogen into 596 million tons of helium every second; the "missing" 4 million tons of mass is converted into the energy that makes the Sun shine. In a sense, the sunlight shining down on us confirms Einstein's famous equation, and because that equation comes straight out of special relativity, sunshine is evidence that relativity is correct.

The second indirect test requires a bit more background. Although we haven't talked about it, a major motivation to Einstein in developing the special theory of relativity was to solve what had seemed to be a problem with the equations of electromagnetism discovered a few decades earlier. Those equations include the speed of light as a constant, but they do not give any indication of the reference frame in which the speed of light should be measured. Before relativity, this seemed like a problem in need of a solution.[2] With relativity, it isn't a problem at all, since relativity tells us that we don't need a reference frame for measuring the speed of light; instead, it is always the same for everyone. More important for our purposes, it turns out that the entire special theory of relativity can be derived from the equations of electromagnetism, but although some physicists had recognized the mathematical ideas (most notably Hendrik Lorentz, for whom the key equations of special relativity are now named as "the Lorentz transformations"), no one before Einstein had truly understood the implications of this fact. Why does this matter? Because those very same equations are what we use to make radios work, as well as virtually every other electrical device we use in the modern world. Every time you turn on your TV, pick up your cell phone, or use your computer, you are confirming the equations of electromagnetism. Because those equations implicitly include special relativity, you are also confirming Einstein's theory.

---

2. The most commonly suggested solution imagined that space was filled with a substance, known as the *ether*, that vibrated as electromagnetic waves passed through it. The 1887 Michelson-Morley experiment was designed to detect this ether, and most scientists were greatly surprised when it did not do so and instead found that the speed of light was always the same.

THE GREAT CONSPIRACY

I'd like to believe that I've made a pretty convincing case for special relativity. I've explained its starting points in the two absolutes, I've led you through a series of thought experiments to see the consequences of those starting points, and I've described the wide range of evidence that supports the whole theory. But how do you know that I'm not making it all up? I could point you to the many other books written about relativity, or I could tell you to talk to other physicists who have studied relativity, but you could always imagine that we're *all* part of a great conspiracy, designed to confuse everyone else so that physicists can take over the world!

Perhaps so, but before you become a conspiracy theorist, you should at least investigate the implications of the conspiracy. So let's pretend for a moment that relativity is *not* true, and that the world behaves the way your old common sense would have expected. This is easy to do. Because all of special relativity flows from the two absolutes—and because the one about the laws of nature being the same for everyone fits in fine with our old common sense—we simply need to get rid of the second absolute. In other words, let's assume that the speed of light is *not* absolute, but instead adds to other speeds just like we expect for balls and cars and airplanes.

Imagine that two cars, both traveling at about 100 kilometers per hour, collide at an intersection, as shown in figure 4.2. You witness the collision from far down one street. If the speed of light were *not* absolute, then the light from each car would come toward you at the normal speed of light *plus* the speed at which the car is coming toward you. For the car coming straight at you, its light would therefore be coming toward you at a speed of $c + 100$ kilometers per hour. The car moving across your line of sight is not coming toward you at all, and therefore its light would just be coming at the normal speed of light, c. As a result, you would in principle see the car coming toward you reach the intersection slightly before the other car.

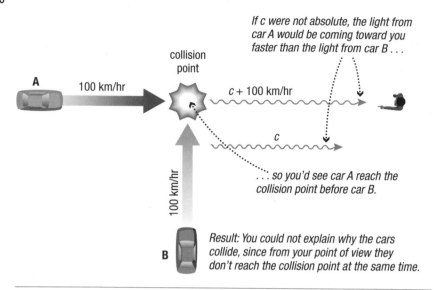

If c were not absolute, the light from car A would be coming toward you faster than the light from car B . . .

collision point

A

100 km/hr

c + 100 km/hr

c

. . . so you'd see car A reach the collision point before car B.

100 km/hr

B

Result: You could not explain why the cars collide, since from your point of view they don't reach the collision point at the same time.

FIGURE 4.2 Two cars collide at an intersection. If the speed of light is absolute, the collision happens unambiguously. But if the speed of light were *not* absolute, different observers would not see the collision unfold in the same way.

So far, this may seem unimportant. After all, the 100-kilometers-per-hour speed of the cars is only about one-millionth of the speed of light, so if you watched from down the street, you'd be hard-pressed to measure the difference in the light arrival times, and the collision would still appear to unfold just as you expect. But what if you observed the collision from very far away? For example, imagine that you used a super telescope to observe the collision from another planet that is 1 million light-years away. Because the light from the first car (the one coming toward you) is coming at a speed one-millionth of the speed of light faster than the light from the second car, over a distance of 1 million light-years this light would end up reaching you *a full year* ahead of the light from the second car.

Think about this. From your point of view, you'd see the first car reach the intersection a full *year* before the second car. This creates a paradox: From the viewpoint of the passengers in the cars, they have collided, yet you saw one car reach the collision point long before the other car had even started its journey! If we live in a universe in which the images we see with light show what really does happen, then the only way to avoid this paradox is to assume that we should *not* have added the car's speed to the speed of light.[3]

You can play with this logic all you want, but the bottom line is pretty clear: In our everyday lives, we assume that light carries images of reality. If two cars collide, we expect all observers to see the collision unfold in the same way, no matter where or how far away they are observing from. But as our thought experiment shows, this agreement on what we see depends on the speed of light being absolute.

So if you're conspiracy minded, you have a choice. You can choose to reject the special theory of relativity, but if you do so, then you must abandon something else your common sense tells you, which is that light shows us how events actually unfold. Alternatively, you can accept the special theory of relativity, with all its mind-bending consequences, but still be safe in the knowledge that it does not contradict anything that your low-speed common sense has ever told you. Given the abundance of evidence that we've discussed for special relativity, it seems like an easy choice.

MAKING SENSE OF RELATIVITY

Assuming that you now accept that relativity is for real, it's time to get back to the issue of making sense of it. Actually, there's not much to do. As we've discussed, relativity does not noticeably affect what we experience in our everyday lives, any more than the real meanings

---

3. I should point out that we would be untroubled by this same paradox if instead of looking at light, we observed some type of particles (e.g., neutrinos) that happened to be emitted by the colliding cars. In that case, we would not be surprised that the event would *appear* to unfold differently from different viewpoints, because we would expect that we'd need to account for the way the particle speeds add. The reason the paradox is troubling with light is because we also expect that light shows us reality; in essence, the paradox shows that we can't have it both ways, with light having an additive speed and showing reality at the same time.

of up and down affect what happens when a child bounces on a bed. So the trick (if there is one) to making sense of relativity must be the same trick that you used to broaden your common sense about up and down after you learned that Earth is round. And that, I'd argue, is just a matter of deciding what's really bothering you.

For most people, the basic ideas of relativity are not too bothersome. After all, why should you be bothered by things like time dilation or mass increase if they're noticeable only at speeds that you've never experienced? The really bothersome part is usually the paradoxes, especially the one we encountered in this chapter, in which you and Al can both claim that the other's time is running slow. Although we've discussed the resolution of that paradox, you're probably still wondering how it can possibly make sense.

To help yourself get past this problem, go outside and answer a simple question. Is the Sun out right now? Let's say that it is, and so you answer "yes." Next, suppose that I decide to argue with you and claim that the Sun is *not* out. At first, you might think I'm nuts. But imagine that you pick up a phone and call me, and I sound very rational as I argue that you're wrong about the Sun being out. Then, you think of a clever idea: You use your phone to take a picture showing that the Sun is indeed out, and send it to me. Just as you're about to declare victory in the argument, I send you a picture with my phone, and it shows clearly that it's nighttime and the Sun is not out.

If you were still a young child, this argument might seem to make no sense at all. However, you now know that it makes perfect sense, at least if you and I happen to be on opposite sides of Earth, so that it is daytime for you and nighttime for me. In other words, we're both talking about the same physical reality—meaning the Sun's actual location in space—but we give different answers about whether or not the Sun is out because we are observing from different places on the round Earth.

In much the same way, the argument between you and Al arises because you are using the old common sense in which we think of space and time as absolutes and expect the speed of light to be relative, meaning that we expect it to add to other speeds just like the speeds of balls and cars. *The theory of relativity tells us that we have it backward*. It is the speed of light that is absolute, and time and space that are relative. Once you accept this simple idea—which is our new common sense—the fact that you and Al can argue about whose time is running slowly is no more surprising than the fact that the two children on opposite sides of Earth can argue about whether it's day or night.

Just as it took you some time as a child to get used to your new common sense about up and down, you may also need a while to get used to your new common sense about time and space. But you now know the goal, which is simply to absorb the implications of the absoluteness of the speed of light, just as you once absorbed the implications of a round Earth. Meanwhile, until you come to accept your new common sense, remember that results are what counts, and every experiment that you perform will agree with every experiment that Al performs. You both live in the same spacetime reality, and you'll both agree on the unambiguous reality of any events that you witness. This is a clear improvement over the paradoxes we encountered when we examined the consequences of a non-absolute speed of light. Special relativity may still seem surprising, but it actually makes the universe make more sense than it did before.

Part 3

EINSTEIN'S GENERAL THEORY OF RELATIVITY

5

NEWTON'S ABSURDITY

IN ADDITION to making many aspects of the universe make more sense than they did before, the special theory of relativity also solved several important and well-known problems in physics, including the apparent problem with the equations of electromagnetism that we discussed earlier. Many physicists had been working on these problems at the time, and several besides Einstein were closing in on the correct solution.

The general theory of relativity is a different story, as most historians of science suspect that without Einstein, it would not have been discovered for many more years. Part of the reason for Einstein's success came from the way in which he approached unsolved problems. Instead of just looking for solutions that worked, he sought solutions that would reveal an underlying simplicity in the universe. In other words, Einstein believed that the universe is inherently simple. It's worth noting that although many scientists share this belief that nature has an underlying simplicity, there's no known scientific reason why it must be so. In that sense, it is a belief that is more like what we usually think of as faith than science. Nevertheless, it remains thoroughly scientific in one key way: If the evidence ever were to show that nature is *not* simple, scientists would revise the belief to fit the new data.

In any event, while most other scientists were satisfied with the special theory of relativity because it solved the well-known problems, Einstein felt that it was not yet complete. He continued his thought experiments and his calculations in search of a way to plug the holes that, to him, still called out for solution. It took him a full decade to work out all the details, with the final result being his 1915 publication of the general theory of relativity.

As it turned out, the general theory not only plugged the holes in special relativity, but it completely redefined the way in which we understand gravity. It is considered Einstein's greatest achievement, and it was what ultimately made him a household name. Interestingly, while many of the predictions of general relativity came as a complete surprise to scientists (and even to Einstein himself), it also solved a few problems that had already been recognized with Newton's older theory of gravity. Indeed, it solved a problem that had greatly troubled none other than Sir Isaac Newton himself.

SPOOKY ACTION AT A DISTANCE

We're so familiar with falling objects and other effects of gravity that it's tempting to think of gravity as a simple idea. But it's not, as you can tell from the embarrassed responses of scientists confronted by kids asking, "But what *is* gravity?" For most of human history, gravity was assumed to be something that operated only on Earth, and the heavens were considered to be a separate and likely unknowable realm. Then, in 1666, a falling apple provided what Newton later said was a moment of inspiration, in which he suddenly realized that the force that held the Moon in orbit around Earth was the same force that made the apple fall to the ground. Not long after, he used the mathematics of calculus—which he invented largely for this purpose—to show that the force of gravity could account for all the known motions of the planets around the Sun.

Newton's universal law of gravitation is a simple equation that allows us to calculate the force of gravity acting between any two objects. It states that the total force depends on the product (multiplication) of the masses of the two objects and on the *inverse square* of the distance between them. In other words, if you triple the distance between two objects, the force of gravity between them goes *down* (the "inverse") by $3^2 = 9$ times (the "square").

By using the gravitational law in conjunction with other ideas, such as his laws of motion, Newton created a *theory* of gravity that successfully explained a broad and diverse array of phenomena, ranging from the reasons why we have weight to why rocks fall and planets orbit. This theory works so well that, at least for most circumstances, there can be little doubt about its validity. Among its most spectacular successes, Newton's theory of gravity was used to predict the existence and location of the planet Neptune before it was discovered by telescope, and it is used to plot the trajectories that take spacecraft to precise landing points on distant worlds.

But if you think about it, there's something very strange about Newton's theory of gravity. Consider Earth's orbit around the Sun. We can easily calculate the force of gravity that holds Earth in its orbit, but how exactly does Earth know that the Sun is there, and hence that it should orbit? After all, Earth does not have senses like sight or hearing, and there is no physical connection holding Earth to the Sun. As expressed in Newton's law, gravity appears to exert what scientists refer to as "action at a distance," operating as though unseen ghosts somehow carry the force instantly across great expanses of space. Newton himself wrote:

That one body may act upon another at a distance through a vacuum . . . and force may be conveyed from one to another, is to me so great an absurdity, that I believe no man, who has . . . a competent faculty in thinking, can ever fall into it.[1]

So now you understand the title of this chapter. "Newton's absurdity" was his very own theory of gravity. As well as it worked, it was

1. Letter from Newton, 1692–1693, as quoted in J. A. Wheeler, *A Journey Into Gravity and Spacetime* (Scientific American Library, 1990).

clear to him that it didn't quite make sense. Oddly, if it bothered others as much, they didn't make a big issue of it during the next two centuries. But you can bet it bothered Einstein. Indeed, when later confronted by the claim from quantum mechanics that a particle in one place can in some cases affect a particle in another place instantaneously (the idea of "quantum entanglement" mentioned briefly in chapter 2), Einstein derisively referred to the claim as "spooky action at a distance." With that in mind, I'm sure you won't be surprised to learn that when Einstein gave us a new view of gravity through general relativity, Newton's absurdity of action at a distance was gone.

SPACE EXPLORERS

My main goal in this chapter and the next is to help you understand the new view of gravity provided by general relativity. As with special relativity, it is an understanding that you must build step by step. To begin, let's consider a simple thought experiment involving explorers.

Imagine that you and everyone around you believe Earth to be flat. As a wealthy patron of the sciences, you decide to sponsor an expedition to the far reaches of the world. You select two fearless explorers and give them careful instructions. Each is to journey along a perfectly straight path, but they are to travel in opposite directions. You provide each with a caravan for land-based travel and boats for water crossings, and you tell each to turn back only after discovering "something extraordinary."

Some time later, the two explorers return. You ask, "Did you discover something extraordinary?" To your surprise, they answer in unison, "Yes, but we both discovered the same thing: We ran into each other, despite having traveled in opposite directions along perfectly straight paths."

Although this discovery would be extraordinarily surprising if you truly believed Earth to be flat, we are not really surprised because we know that Earth is round. As shown in figure 5.1, the "straight" lines

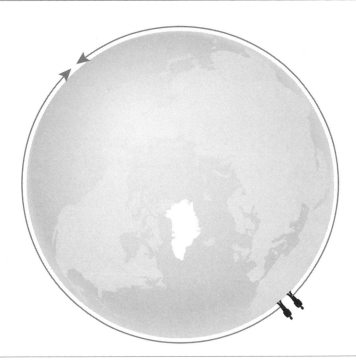

FIGURE 5.1 On Earth, two travelers who start out going "straight" in opposite directions will meet on the opposite side of the globe. We're not surprised, because we recognize that it is due to the curvature of Earth's surface.

followed by the explorers actually curve around the globe, causing them to meet up. In a sense, the explorers followed the *straightest possible* paths, but the curvature of Earth's surface means that these paths were curved.

Now let's consider a somewhat more modern scenario. You are floating freely in a spaceship somewhere out in space. Hoping to learn more about space in your vicinity, you send two explorers out in space probes going in opposite directions. Aside from a brief push to get each of them started, they do not use any engines, so they go perfectly straight away from you. Imagine that, some time later, you get radio messages from both explorers, telling you that they have just passed each other! How can this be, when they went in opposite directions when they left you?

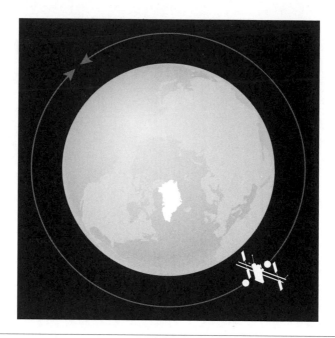

FIGURE 5.2 Two space probes launched in opposite directions in Earth orbit will meet in a way that looks almost identical to the explorers in figure 5.1, yet we usually give a completely different reason for the meeting of the probes, attributing it to the "action at a distance" of Earth's gravity. But could the real reason be that *space* is somehow curved in much the same way that Earth's surface is curved?

In fact, it can occur quite naturally if you happen to be orbiting Earth while you float freely in your spaceship. For example, as figure 5.2 shows, two probes sent out in opposite directions from a space station would meet up on the opposite side of Earth. Since the time of Newton, we've generally explained the curved paths of the probes as an effect caused by the "action at a distance" of Earth's gravity. But notice the similarity of the paths in figures 5.1 and 5.2. By analogy with the explorers journeying in opposite directions on Earth, might we instead conclude that the probes meet because *space* is somehow curved? That idea lies at the heart of Einstein's general theory of relativity. But before it will make sense to you, we need to go back to thinking more deeply about the relativity of motion.

IS THE RELATIVE IN RELATIVITY ALWAYS RELATIVE?

Recall that the theory of relativity gets its name from the idea that all motion is relative. As we demonstrated with the airplane on the Nairobi-to-Quito flight (see figure 2.1), there is no absolute answer to the question of "who is really moving?" All we can say is that the airplane moves *relative* to Earth, but observers in different reference frames will view that relative motion differently.

The idea that motion is relative is very simple, and we've seen how it comes into play with our thought experiments in free-float reference frames. When you and Al both float freely in your spaceships, you can each legitimately claim to be the one at rest, saying that it is the other who is in motion. But what if one of you is *not* floating freely; can you both still claim to be at rest? Let's investigate.

Imagine that you and Al are both floating freely in space when you suddenly decide to fire your rocket engines with enough thrust to give you a continuous acceleration of "1g," which is the acceleration with which objects fall toward the ground on Earth. (Numerically, 1g is equal to 9.8 meters per second squared.) As long as you keep firing your engines, Al will see you accelerating away, with your speed growing ever faster. From his point of view, he's still floating stationary in his ship, so he sends you a radio message saying, "Good-bye, and have a nice trip!"

If *all* motion is relative, you should be free to claim that *you* are the one who is stationary, and that it is Al who is receding into the distance at ever-faster speeds. You might therefore wish to reply, "Thanks, but I'm not going anywhere. You're the one who is accelerating away."

However, the firing of your engines has introduced a new element that was not present in our earlier thought experiments: As shown in figure 5.3, the force created by the engines will push you against the floor of your spaceship, meaning that you will no longer be weightless. In fact, because the engines are giving you an acceleration of 1g, the force will allow you to walk on the floor with your normal Earth weight. Therefore, if Al looks into your spaceship with a telescope,

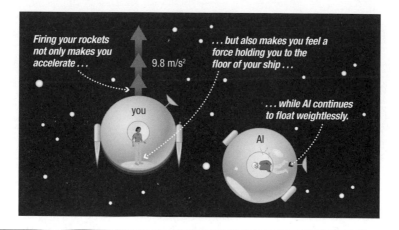

FIGURE 5.3 When you and Al were both floating weightlessly, it was clear that you could each claim to be the one at rest while the other is moving. But when you fire your spaceship's engines, they make you accelerate (which means an ever-increasing speed) *and* generate a force that gives you weight with which to stand on the floor. So how can you still claim to be at rest?

he may respond back: "Oh, yeah? If you're not going anywhere, why are you stuck to the floor of your spaceship, and why do you have your engines turned on? And if I'm accelerating as you claim, why am I weightless?"

You must admit that Al is asking very good questions. It certainly *looks* as if you really are the one who is accelerating, in which case you cannot legitimately claim to be stationary, and the fact that Al is still floating freely seems inconsistent with your claim that he is accelerating, since acceleration should be accompanied by a force. At first glance, it seems that motion is no longer relative when we introduce acceleration.

This idea did not sit well with Einstein. It seemed to him that *all* motion should be relative, regardless of whether any acceleration is involved. If we apply this notion to our thought experiment, it means we need some way to explain the force that you feel due to firing your rocket engines without necessarily assuming that you are accelerating through space, while simultaneously explaining Al's weightlessness even while you claim that he is accelerating away from you.

EINSTEIN'S HAPPIEST THOUGHT

In 1907, just two years after finishing special relativity, Einstein hit upon what he later called "the happiest thought of my life." To understand his happy thought, put yourself back in your accelerating spaceship. As it accelerates through space at 1g, you will be able to sit down, get up, or walk around on the floor with your normal Earth weight. If you toss a ball in the air, from your point of view it will fall back down just as it would on Earth. In fact, if you close all the window shades, everything inside your spaceship will seem to behave just as it would at home in your room on Earth (figure 5.4).

If you were a physicist living at the time, your first reaction to Einstein's happy thought might have been "Well, duh." Since the time of Newton, it had been well known that the effects of gravity and the effects of acceleration would feel the same. But to those scientists who thought more deeply about it (and they were many), this seemed to be a rather astonishing coincidence. Consider Galileo's famous discovery that all objects fall with the same acceleration on Earth (disregarding air resistance). If you apply any other type of force to

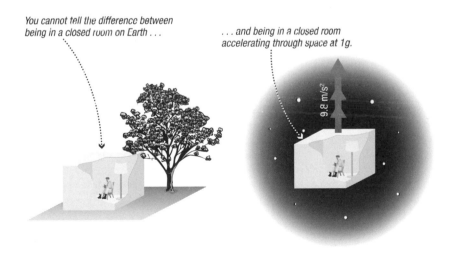

You cannot tell the difference between being in a closed room on Earth . . .

. . . and being in a closed room accelerating through space at 1g.

9.8 m/s²

FIGURE 5.4 The effects of acceleration feel the same as the effects of gravity.

objects of different masses, such as the force you apply when trying to throw them, it's more difficult to accelerate a larger mass than a smaller mass; that is why it's harder to throw the metal ball used in shot put than a baseball. Yet when we deal with gravity, the acceleration comes out exactly the same, regardless of the mass.[2]

From a pre-relativity point of view, it was almost as if nature was showing us two boxes, one labeled "effects of gravity" and the other labeled "effects of acceleration." Scientists shook, weighed, and kicked the boxes but could never find any obvious differences between them. They concluded: "What a strange coincidence! The boxes seem the same from the outside even though they contain different things." Einstein's revelation was, in essence, to look at the boxes and say that it is not a coincidence at all. Rather, he said that the boxes appear the same from the outside *because they contain the same thing.*

This remarkable idea is called the *equivalence principle.* Stated more precisely, it says: The effects of gravity are exactly equivalent to the effects of acceleration.[3] With this principle, the relativity of all motion is restored, as we can see by going back to Al's questions as he watched you accelerate away with your engines firing.

His first question was how you could claim to be at rest when you are no longer floating freely but instead are feeling a force that gives you weight on the floor of your spaceship. With the equivalence principle, you can claim that you feel weight because of gravity. That is, you can

---

2. Mathematically, this coincidence arises when we apply Newton's second law, force = mass × acceleration. Most forces do not depend on an object's mass; for example, the electromagnetic force depends on charge, which is unrelated to mass. But when we use gravity as the force in Newton's second law, an object's mass appears on both sides of the equation and therefore cancels, so that the object's acceleration does not depend on its mass. For this reason, the coincidence is sometimes said to be the fact that the "gravitational mass" (the mass that appears in the force of gravity on the left side of the equation for Newton's second law) is equal to the "inertial mass" (the mass that appears next to acceleration on the right side). Before Einstein, there was no known reason why the gravitational mass and inertial mass should have the same value.

3. Technically, this equivalence holds only within small regions of space. Over larger regions, the gravity of a massive object, such as a planet, varies in ways that would not occur due to acceleration; for example, such variation explains why gravity gives rise to tidal forces that do not occur in an accelerating spaceship.

claim that the space around you is filled with a gravitational field point-ing "downward" toward the floor of your spaceship.

Al's other questions concerned why you would be firing your engines if you were stationary, and why he would be weightless if, as you claimed, he was accelerating away from you. Both are now answered easily. Al is weightless because he is in freefall through the gravitational field, and anyone in freefall is always weightless.[4] Your engines are fir-ing to *prevent* you from falling in the same way.

To summarize, the equivalence principle allows you to claim that the situation is equivalent to what it would be if you were hovering over a cliff while Al had fallen over the edge (figure 5.5). That is, you can respond: "Sorry, Al, but I still say that you have it backward. I'm using my engines to prevent my spaceship from falling, and I feel weight because of gravity. You're weightless because you're in freefall. I hope you won't be hurt by hitting whatever lies at the bottom of this gravi-tational field!"

Of course, when you look at figure 5.5, there may still seem to be a problem: Where's the cliff, or the planet producing the gravity that you claim to feel? More generally, while it's easy to *say* that the effects of gravity and acceleration are equivalent, they sure don't seem to look the same. After all, it would seem difficult to confuse a person standing on Earth's surface with one hurtling through space at ever-increasing speed.

This difference between the normal appearances of gravity and accel-eration takes us to the heart of Einstein's happiest thought. Einstein didn't just say that the effects of gravity and acceleration *feel* the same; as we discussed, everyone already knew that. He said that they *are* the same. Therefore, according to Einstein, if they *look* different, it must be because we're not seeing the whole picture. What part of the picture are we missing? Once again, it's that fourth dimension of spacetime.

---

4. In case you are wondering why freefall means weightlessness, imagine standing on a scale on a high platform. As long as the platform remains intact, your feet will be pushing on the scale, which will therefore read your normal weight. But if the platform breaks, sending you and the scale plummeting down in freefall, your feet will no longer be pushing on the scale, which means the scale will read zero—that is, you will have become weightless.

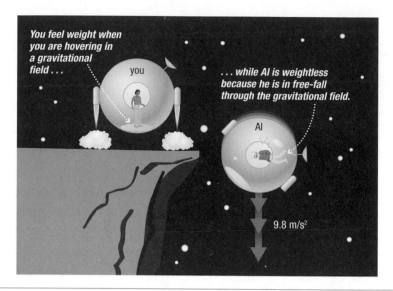

FIGURE 5.5 Imagine using your engines so that you can hover over a cliff, while Al drops past in free fall. You will be stationary and feeling weight due to gravity, while Al is weightless as he accelerates downward. According to the equivalence principle, you can claim that this is the situation *even if there's no planet or cliff around*.

Recall that different observers can measure time and space differently, but spacetime is the same for everyone. In much the same way, different observers can have different perceptions about gravity and acceleration, but in spacetime we'd find that they both look the same.

EQUIVALENCE IN SPACETIME

We now come to the key to general relativity, which is to understand how effects of gravity and acceleration can look the same in four-dimensional spacetime. To do this, we need a way to visualize the different types of paths that objects can have through spacetime. Let's start with a simple example.

Suppose you drive your car along a straight road from home to work as shown in the top panel of figure 5.6. At 8:00 a.m., you leave your house and accelerate to a speed of 60 kilometers per hour. You maintain this speed until you come to a red light, where you decelerate to a stop. After the light turns green, you accelerate again to 60 kilometers per hour, which you maintain until you slow to a stop when you reach work at 8:10.

What does your trip look like in spacetime? If we could see all four dimensions of spacetime, we'd see the three dimensions of your car tracing a path stretched out through the 10 minutes of time taken for your trip. We can't visualize all four dimensions at once, but in this case we have a special situation: Your trip progressed along only one dimension of space because you took a straight road. Therefore, as shown in the bottom panel of figure 5.6, we can represent your trip in spacetime by drawing a graph with one dimension of space on the horizontal axis and time on the vertical axis. This type of graph is called a *spacetime diagram*, and an object's path through spacetime is called its *worldline*.

Our spacetime diagram for your car trip reveals three crucial features of worldlines: (1) When an object is stopped (stationary) in your reference frame, its worldline is vertical; that is, it does not move in space but goes straight up through time. (2) When an object moves relative to you at constant velocity, its worldline is straight but slanted, because it moves the same amount of distance in each unit of time. (3) When an object accelerates or decelerates, its worldline is curved, because the amount of distance it moves is changing with each passing second.

We can use these ideas to explore the relativity of motion. We'll start with our special relativity thought experiments, in which you and Al both float freely while seeing each other in relative motion. The left side of figure 5.7 shows how *you* would draw a spacetime diagram for the situation. You consider yourself to be at rest, so your own worldline is vertical, while Al's worldline is straight and slanted because he is moving by you at constant velocity. The right side of the figure shows Al's version of the spacetime diagram, in which he has his own worldline vertical and yours slanted. The fact that the two graphs

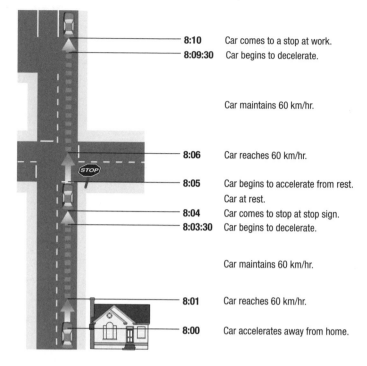

8:10 Car comes to a stop at work.
8:09:30 Car begins to decelerate.

Car maintains 60 km/hr.

8:06 Car reaches 60 km/hr.

8:05 Car begins to accelerate from rest.
Car at rest.
8:04 Car comes to stop at stop sign.
8:03:30 Car begins to decelerate.

Car maintains 60 km/hr.

8:01 Car reaches 60 km/hr.

8:00 Car accelerates away from home.

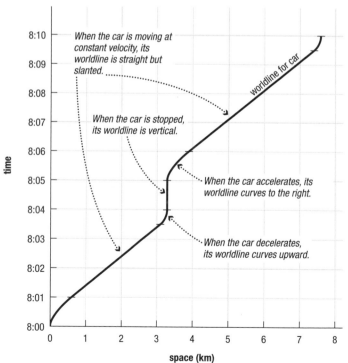

When the car is moving at constant velocity, its worldline is straight but slanted.

When the car is stopped, its worldline is vertical.

worldline for car

When the car accelerates, its worldline curves to the right.

When the car decelerates, its worldline curves upward.

time

space (km)

have their space and time axes in different orientations relative to the two worldlines (yours and Al's) explains why your measurements of space are different from Al's measurements of space, and why your measurements of time are different from his measurements of time. But notice that if you ignore the axes, which only represent arbitrarily chosen coordinate systems, the two diagrams actually are identical: you can turn one into the other simply by rotating the page a bit as you look at them. The fact that the two diagrams are identical means that the spacetime reality is the same for both of you.

Let's now return to our thought experiment from this chapter, in which you are firing your engines so that Al sees you accelerating away from him. From Al's point of view, his own worldline looks just the same as it did before. But in this case, your acceleration means that Al will claim you have a curved worldline, as shown on the left of figure 5.8. Now we turn to your version of the spacetime diagram. Since you are claiming to be at rest in a gravitational field, you might wish to draw your worldline as vertical, just as it was when you were in a free-float frame. So go ahead and draw a straight line on a piece of paper, like the one shown on the right of figure 5.8.

Here is your task: According to what we learned in special relativity, there is only one spacetime reality, and everyone must agree upon it. Therefore, in spacetime, you and Al must both agree on the shape of your worldline. This was easy for constant velocity (like that in figure 5.7), because you and Al agreed that your worldlines were both straight; you were simply looking at them with different coordinate axes. But how can you both agree on the shape of your worldline when he says it is curved while you choose to draw it as straight? The answer is—drumroll, please!—*bend your piece of paper.*

FIGURE 5.6 (*Top*) This diagram shows a car trip from home to work along a straight road, indicating all the places and times at which the car's motion changed. (*Bottom*) The same trip represented on a spacetime diagram, with space on the horizontal axis and time on the vertical axis.

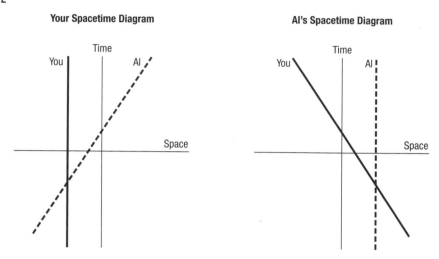

Your Spacetime Diagram

Al's Spacetime Diagram

FIGURE 5.7 When you and Al are moving relative to each other but are both floating freely in your spaceships, Al's spacetime diagram is just a slightly rotated version of yours. The fact that you both agree on the relative placement of the worldlines reflects the spacetime reality, while the fact that you each place the time and space axes differently explains why your measurements of space and time will differ.

This answer is so deceptively simple that I'll go over it again for you in a different way. The key point is that there is only one spacetime reality. When we deal with straight lines, they're really all the same, except rotated from one another. But curves and straight lines are different. If your worldline is curved in spacetime, it is curved, and that's that. Einstein's revelation, his happiest thought, in essence comes down to the idea that there are two ways to get a curved worldline: You can draw it as a curve, which is what Al did on his spacetime diagram, or you can draw it as a "straight" line on a curved piece of paper. Its final shape is curved either way.

We are finally ready to put this in terms of the equivalence principle. From Al's point of view, you have a curved worldline because you are accelerating through a spacetime that we've represented with a flat piece of paper. From your point of view, you are stationary in

Al's Spacetime Diagram

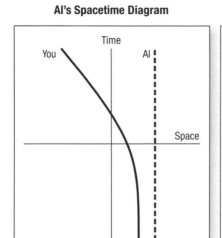

Your View of Your Worldline

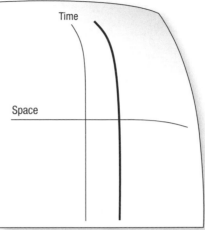

FIGURE 5.8 (*Left*) When you are accelerating through space, your worldline will be a curve when Al draws it on his spacetime diagram. (*Right*) According to the equivalence principle, you can claim that you are at rest in a gravitational field, which might lead you to draw your own worldline as straight. How, then, can you and Al agree on the spacetime reality?

a gravitational field, but that gravitational field is causing your piece of paper (representing spacetime) to curve. In other words, the reason that the effects of acceleration and the effects of gravity are the same is because acceleration and gravity are simply alternate ways of describing a curved path through spacetime. Or, to pull out the critical conclusion: *Gravity arises from curvature of spacetime.*

Before we go on, I need to give you one small warning: While the bent piece of paper analogy is very useful, it is not a perfect representation of spacetime. As I mentioned in chapter 4, the actual geometry of spacetime is more complex than the geometry we learn in high school, and you can generate some misconceptions if you take the bent paper analogy too far. Rest assured that there *are* ways to model spacetime more accurately, but they require mathematical techniques (many developed by Einstein) that are beyond the scope of this book.

GHOSTBUSTERS

There's no way to overstate the significance of what we have just discussed. Through the equivalence principle, Einstein has given us an entirely new way to view gravity.

Flip back a few pages and look again at the two space probes that meet up in figure 5.2. Imagine that you photocopied and enlarged that figure, then pasted it down in a big, round salad bowl. The two space probe paths would still start out in opposite directions around the bowl, but this time they would meet up not because you drew circles, but because the *shape of the bowl left them no choice.* Spacetime is not really shaped quite like a four-dimensional salad bowl, but the basic idea is still the same. That is, Einstein's new view of gravity says that the two probes meet up in orbit for the same reason that the two explorers meet up on Earth: In both cases, they are going as straight as they possibly can, but they are constrained by the geometry of the space in which they move.

It's worth a moment to explore this idea a bit further. Although Earth itself is a three-dimensional object, Earth's surface is only two-dimensional, because there are only two independent directions of motion possible: north-south and east-west. It is this two-dimensionality that once made people think that Earth was flat, but we didn't have to see Earth from space to learn otherwise. In the real-life version of the explorer story from earlier in this chapter, our ancestors learned that Earth's surface is curved by studying the observations made by ancient explorers.

Before general relativity, we naively thought that space was just as "flat" as our ancestors once thought Earth's surface to be. Thanks to Einstein, we now know that the probes are teaching us about the true shape of space (and spacetime) in the same way that ancient explorers taught us about Earth's true shape. In this case, the fact that the probes meet up in orbit tells us that the space around Earth must be curved in a way that makes this meeting a natural outcome for their "straight" paths. The fact that we cannot *see* this curvature does not matter; we are able to measure it by observing the orbital paths.[5] Moreover, even though we know that the reason for the curvature is the "gravity" of Earth, the probes needn't care whether Earth is there or not. They are simply following the pathways allowed by the *local* structure of the space in which they move.

In the old Newtonian view, gravity was a force that exerted "action at a distance" between two objects. With his general theory of relativity, Einstein has removed the mystery of this action, vanquished any spooky ghosts that may have been responsible for it, and cleared up Newton's absurdity by showing that gravity arises as a natural consequence of curvature of spacetime. Orbits are no longer a result of a mysterious gravitational force, but just the straightest possible paths through curved regions of spacetime.

---

5. By analogy to the fact that Earth's two-dimensional surface curves through three-dimensional space, it's natural to wonder what "other" dimensions three-dimensional space (and four-dimensional spacetime) are curved through. The somewhat unsatisfying answer is that if such dimensions exist at all, they probably have no more bearing on us than the three-dimensionality of Earth has on an ant crawling along the surface. That is, we can work mathematically with a curved four-dimensional space without invoking or knowing anything about any other dimensions. Incidentally, for readers familiar with the way some modern physics theories propose additional dimensions wrapped up on a subatomic level, those are *not* thought to be related to whatever "other" dimensions might exist beyond spacetime.

6

REDEFINING GRAVITY

THE IDEA of gravity arising from curvature of spacetime takes some getting used to, especially since the only visualizations we can do are somewhat inadequate two-dimensional analogies, such as bent pieces of paper or orbital paths in salad bowls. Nevertheless, the appeal of this new idea should be clear. Just as special relativity made the universe make more sense than it did before, so does general relativity. As we saw in chapter 5, general relativity eliminates Newton's absurdity, it allows us to treat all motion as relative (or, more precisely, to get the same answers no matter what reference frame we choose, an idea in relativity called "general covariance"), and it explains what had seemed to be a surprising coincidence between the effects of gravity and the effects of acceleration as a natural outcome of an underlying simplicity in nature.

From the vantage point of history, the most difficult task that Einstein faced after developing the equivalence principle was finding a way to get other scientists to accept it. After all, the history of science is littered with nice ideas that sounded appealing at the time, and the mere fact that the equivalence principle was Einstein's "happiest thought" was not enough to give it scientific legitimacy. What Einstein needed was a complete, mathematically grounded description of gravity based on the equivalence principle. Moreover, he needed to show that his theory would make at least some quantitative predictions that were different from the predictions made by Newton's theory of gravity, so that actual observations or experiments could test whether his new theory worked better than the old one.

This need for mathematical precision largely explains why it took Einstein eight years from the time he first thought of the equivalence principle to his publication of the general theory of relativity. The major problem was that making calculations based on the equivalence principle turned out to be much more mathematically demanding than the calculations he had needed for special relativity. In particular, while all the major consequences of special relativity can be derived with simple algebra, doing calculations for curved, four-dimensional spacetime requires esoteric branches of mathematics that had not been fully explored before Einstein. Indeed, in much the same way that Newton had to invent calculus to work out his theory of gravity, the calculations of general relativity required the invention of new mathematical techniques.

We won't deal with the mathematics of general relativity in this book, but there are at least three reasons why I think it is important to be aware that this mathematics exists. First, our analogies will not be perfect, but any concerns you might have about them should be alleviated by knowing that the actual theory rests on a solid mathematical foundation. Second, I have found that many people underestimate the importance of mathematics to science; while science is based on ideas, the requirement that ideas be testable almost always means that you have to find mathematical ways to calculate with them. Third, I hope that some young readers will be inspired to go beyond what we cover in this book, and I want you to be aware that this means you'll need to focus with particular diligence on the study of mathematics.

With that said, it's time for us to look deeper into the significance of the statement that "gravity arises from curvature of spacetime." We'll start by considering the conditions in spacetime that allow for weightlessness, and then use our insight to figure out the underlying cause of spacetime curvature.

STRAIGHTEST POSSIBLE PATHS

Recall that you and Al were both weightless when you were moving at constant velocity relative to one another in deep space, and the symmetry of your conditions made it easy for you each to claim that you were the one at rest. When you turned on your engines, however, that symmetry was broken because you felt weight while Al remained weightless. In order to continue claiming that you were at rest, you had to invoke the equivalence principle, saying that you felt weight due to gravity and that Al was weightless because he was in freefall. Al, of course, continued to claim that his weightlessness was the result of being at rest in deep space.

Notice that you and Al both agree that he is weightless; you differ only in your explanations of it. According to the equivalence principle, both of your viewpoints must be expressing the same spacetime reality, which means that the spacetime path of an object floating in deep space must be the same as that of an object in freefall. Now, much as we did in finding out how gravity and acceleration could appear the same in spacetime, we must ask what is "the same" about the paths of objects in deep space and of objects in freefall.

Perhaps surprisingly, we can get the answer by thinking about one more situation in which objects are weightless, which is when they are in orbit. The reason that astronauts aboard the International Space Station are weightless is that they are in a continual state of freefall toward Earth. You can see why by imagining a very tall tower (figure 6.1). If you simply step off the tower, you will fall straight downward, but if you run and jump, you'll land a short distance away. The faster you run, the farther you'll go before landing. If you could somehow run fast enough—about 28,000 kilometers per hour (17,000 miles per hour) at the altitude of the Space Station—a very interesting thing would happen: By the time gravity had pulled you downward as far as the length of the tower, you would already have moved far enough around Earth that you'd no longer be going down at all. Instead, you'd be just as high above Earth as you'd been all along,

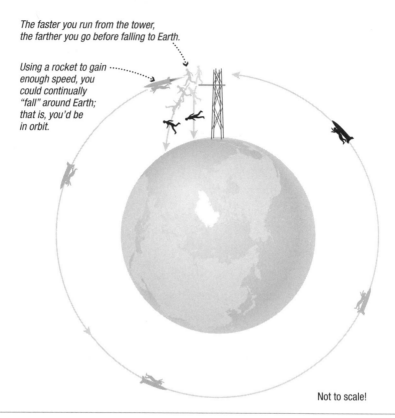

The faster you run from the tower, the farther you go before falling to Earth.

Using a rocket to gain enough speed, you could continually "fall" around Earth; that is, you'd be in orbit.

Not to scale!

FIGURE 6.1 This diagram explains why orbit represents a state of continual freefall. Adapted from a similar illustration in Jeffrey Bennett, Megan Donahue, Nick Schneider, and Mark Voit, *The Cosmic Perspective*, 7th ed. (2014). By permission of Pearson Education, Inc., Upper Saddle River, N.J., which in turn was adapted from a diagram in *Space Station Science*, by Marianne Dyson.

but a good portion of the way around the world. In other words, if you move fast enough, you can continue to "fall around" Earth forever, which is the same thing as being in orbit.

Now, recall from the last chapter that orbiting objects, somewhat like marbles in a salad bowl, are following paths through curved spacetime that are as "straight" as the local geometry allows; we refer to such paths as the "straightest possible," since the local geometry prevents them from being truly straight. Because all freefall trajectories must be equivalent in spacetime, we conclude that *all* such trajectories

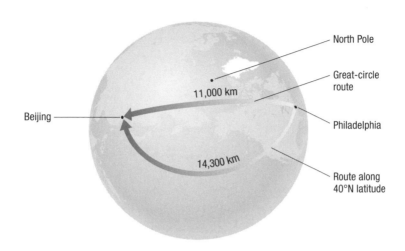

FIGURE 6.2 There are many possible paths between Philadelphia and Beijing, but the great circle route is the shortest and straightest. All other routes, such as the one that stays at constant latitude, are longer and "more curved."

must represent objects that are following straightest possible paths between two points in spacetime. In other words, because the equivalence principle tells us that floating in deep space is equivalent to free-fall, the common spacetime characteristic of any path on which you are weightless is that it is the *straightest possible* path. Whenever you feel weight—such as when you are firing your rocket engines or standing on Earth—you are *not* on the straightest possible path.

An analogy may help summarize what we've found. As shown in figure 6.2, the cities of Beijing and Philadelphia are both at latitude 40°N but are almost halfway around the world from each other in longitude. Therefore, the shortest and straightest possible path between them is a "great circle" route (meaning a route that would cut Earth in half if we extended it) that crosses nearly over the North Pole. There are many other possible paths between the cities, but every other path will be longer and "more curved" than the straightest possible path. In the same way, there are many possible paths between any two particular points in spacetime. But only one path is the straightest possible, and that is the only one on which you will be weightless.

A NEW VIEW OF GRAVITY

The fact that orbits represent straightest possible paths through spacetime is very useful: It means that even though we can't *see* the curvature of spacetime, we can map it out by observing orbital paths. We've already done this for one case, in which we used the orbits of the probes to tell us that space must have a curvature in the vicinity of Earth that causes the probes to go round and round in orbit.

We can take this idea further by mapping many orbits. For example, objects orbiting closer to Earth trace out tighter ellipses than orbits of objects higher up, which tells us that space must get more curved as you get closer to Earth. Similarly, an object orbiting a more massive planet, like Jupiter, orbits faster than an object orbiting at the same distance from Earth. This tells us that space must be more highly curved (to cause the higher speed) around Jupiter than around Earth. These ideas lead to a crucial conclusion: *The curvature of spacetime is shaped by the masses within it.* The greater the mass, the more it curves spacetime around it. A small object orbiting a more massive object simply follows the straightest possible path that it can given the local structure of spacetime.

A common way to visualize this idea is by representing spacetime with a stretched rubber sheet on which we place masses to represent objects like stars and planets.[1] Figure 6.3 shows a rubber sheet model of spacetime around the Sun. We represent the Sun with a heavy mass placed on the rubber sheet, and we can think of the planets as marbles that are circling in the depression created by the heavy mass. In other

---

1. More technically, these so-called *embedding diagrams* represent the shape of a two-dimensional slice through an object or region of space as it would appear if we could see it in a multidimensional hyperspace. Figure 6.3, for example, shows what to us would appear to be a flat plane that cuts through the Sun's equator and the orbits of the planets (which are approximately all in the same plane), but which would appear to a multidimensional being as a curved surface with a shape approximately as shown.

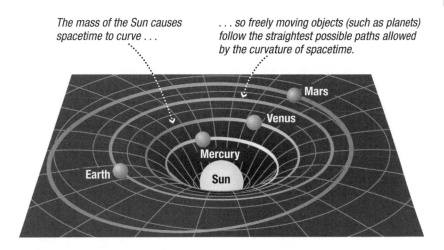

The mass of the Sun causes spacetime to curve . . .

. . . so freely moving objects (such as planets) follow the straightest possible paths allowed by the curvature of spacetime.

Mars

Venus

Mercury

Earth

Sun

FIGURE 6.3 According to general relativity, planets orbit the Sun in much the same way that marbles would go around a stretched rubber sheet: Each planet is going as straight as it can, but the curvature of spacetime causes its path through space to curve.

words, the central mass causes the curvature of the rubber sheet in much the same way that the Sun's mass causes the curvature of spacetime. Similarly, the marbles orbit around because they are following the straightest possible paths in their vicinity of the rubber sheet, much as the planets orbit the Sun because they are following the straightest possible paths in the local spacetime.

As with our other spacetime analogies, keep in mind that the rubber sheet analogy should not be taken too far. It does a reasonable job of representing how orbits work, but a rubber sheet's geometry is not a perfect match to spacetime geometry; in fact, the rubber sheet does not show the *time* part of spacetime at all. Most important, remember that the rubber sheet is a two-dimensional analogy to a four-dimensional reality, and none of the distortions that we see on the rubber sheet are actually visible when we look out into space. When you look through a telescope, the Sun and planets appear simply as spheres, not as spheres sitting on top of rubber sheets or bowls.

GRAVITATIONAL LENSING

With our new view of gravity, we are ready to look for testable consequences of Einstein's theory. We'll begin by seeking observable consequences of the curvature of spacetime around massive objects.

We cannot directly see spacetime curvature, but we can probe it by observing the paths of light rays. Light always travels at the same speed, which means it never accelerates or decelerates, so light must always follow the straightest possible path through space and spacetime. If space itself is curved, then light paths will curve as they pass through that space. Einstein recognized this fact, and it led him to make one of the most remarkable predictions in the history of science: He predicted that stars should appear slightly out of position when viewed near the Sun during a total solar eclipse.

Astronomers can measure the positions and angular separations of stars quite accurately in the night sky. But suppose that we look at two stars during the daytime—let's call them Star A and Star B—and that Star A appears closer to the Sun in the sky. As figure 6.4 shows, the fact that space is more curved near the Sun should cause the light from Star A to follow a more curved path than the light from Star B, with the measurable consequence being that the angular separation of the two stars should appear smaller than it does when we view them at night.[2]

Using Einstein's prediction, two teams of astronomers set out to observe the positions of stars during the total eclipse of May 29, 1919. An expedition led by Arthur Eddington viewed the eclipse from Principe Island in the Gulf of Guinea off the western coast of Africa, while Andrew Crommelin led an expedition to view the eclipse in northern Brazil. The results, announced on November 6

---

2. Interestingly, Newton's theory of gravity also predicted bending of starlight around the Sun (essentially because light is treated the same as a particle with mass moving at the speed of light), a prediction that scientists were well aware of at the time. However, Einstein found that general relativity predicted twice as much bending as Newton's theory, making it possible to test which theory agreed with observations.

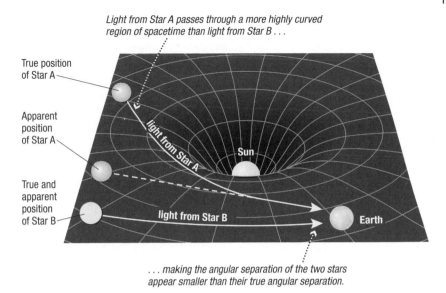

Light from Star A passes through a more highly curved region of spacetime than light from Star B . . .

True position of Star A

Apparent position of Star A

True and apparent position of Star B

light from Star A

Sun

light from Star B

Earth

. . . making the angular separation of the two stars appear smaller than their true angular separation.

FIGURE 6.4 If we observe stars in the daytime, as we can during a solar eclipse, the curvature of space near the Sun can cause measureable changes in the positions of stars.

of the same year, were a resounding success for Einstein. The fact that scientists had observationally confirmed curvature of spacetime generated intense media attention, and Einstein, who had been little known outside the scientific community, suddenly became almost a household name.

The bending of light by gravity, usually called *gravitational lensing* by analogy to the way glass lenses bend light, has since been confirmed many more times. Astronomers have continued to observe stars during subsequent eclipses. By the 1960s, the advent of radio telescopes allowed astronomers to measure daytime star positions even without an eclipse, since sunlight does not interfere with radio observations the way it does with visible light. Today, the deflection of starlight by the Sun has been measured with great precision, and Einstein's predictions agree with the measurements to better than 1 part in 10,000. In other words, to the accuracy of today's technology, general relativity gives an exact match for the observed deflection of starlight. (The European Space Agency's *Gaia* mission, scheduled for launch shortly after this book went to press, should be able to test this prediction even more stringently, to about 2 parts in 1,000,000.)

Looking beyond our solar system, gravitational lensing can also distort the light of distant objects, often with spectacular effect. Figure 6.5 shows how it works. When a distant star or galaxy lies behind another massive object (as seen from Earth), the intervening object curves spacetime in its vicinity, so that light paths that would have gone off in different directions can end up converging at Earth. Depending on the precise four-dimensional geometry of spacetime between us and the observed star or galaxy, the image we see may be magnified or distorted into arcs, rings, or multiple images of the same object. Figure 6.6 shows the effect in an amazing image from the Hubble Space Telescope.

It's worth noting that gravitational lensing not only makes pretty pictures but is also useful. Astronomers have found so many cases of gravitational lensing in the distant cosmos that they now routinely use it "in reverse" to map out the distribution of *dark matter* in the universe. As you may have heard, strong evidence suggests that most of the matter in the universe gives off no light at all (hence the name *dark matter*), which means it is invisible to all of our telescopes. However, astronomers can use the distortions of light caused by gravitational lensing to calculate the distribution of mass that must be causing it. Because mass has the same effect whether it is ordinary matter or dark matter, these calculations can be used to figure out where dark matter is located, and how much of it there is.

An interesting side note to Einstein's successful prediction of gravitational lensing comes from his own words about the observational tests. Remember that Einstein looked for theories that, to him, had a sense of beauty and consistency and that ensured that the laws of physics would be the same in all reference frames. He was convinced that general relativity provided a much more beautiful and sensible view of the universe than did the old view of gravity that it replaced.

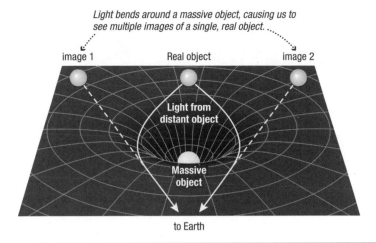

Light bends around a massive object, causing us to see multiple images of a single, real object.

image 1   Real object   image 2

Light from distant object

Massive object

to Earth

FIGURE 6.5 This diagram shows how gravitational lensing can cause us to see two distinct images of a single real object. Note that if you look at the same idea in three dimensions (rather than the two shown here), you can get many more images, or even arcs or rings. If the separation between the images is small, the effect will be to magnify a single image.

Distorted images of background galaxies

FIGURE 6.6 This Hubble Space Telescope image shows a cluster of galaxies known as Abell 2218. The thin arcs are caused by gravitational lensing, as the cluster's gravity distorts the light from galaxies that lie behind it. NASA/Hubble Space Telescope Science.

So when a student in 1919 asked him how we would have reacted if the eclipse observations had *not* confirmed his theory's prediction, he is said to have replied: "Then I would feel sorry for the good Lord. The theory is correct anyway." It's not the most scientifically valid statement he could have made, since observations and experiments are the supreme tests in science, but it's illustrative of the importance that Einstein placed on having a universe that makes sense.

GRAVITATIONAL TIME DILATION AND REDSHIFT

Recall that during your black hole trip in chapter 1, you sent a clock falling toward the black hole and noticed its time running slower and its numerals reddening in color as it fell. Both observations are consequences of a startling prediction that general relativity makes about gravity and time. As usual, we'll use a thought experiment to understand the prediction.

Imagine that you and Al, instead of being in two separate spaceships, are in the same spaceship—one that is long and rocket-shaped, with the two of you at opposite ends (figure 6.7). The ship is in space with engines off, so that you and Al are both floating weightlessly, and you each have a light that flashes once per second. Because you are both floating freely with no relative motion between you, you are both in the same reference frame. Therefore, you will see each other's lights flashing at the same rate.

Now consider what happens when you fire the spaceship engines. Because both of you are in the ship, you and Al will both feel weight pressing you to the floor, Al on the lower deck (at the back of the ship) and you on the upper deck (at the front of the ship). You can explain this weight as being due to either acceleration or gravity, but let's assume that you choose to call it acceleration, perhaps because you notice your speed increasing relative to some nearby planet. What does the acceleration do to the way you see the flashes from each other's lights?

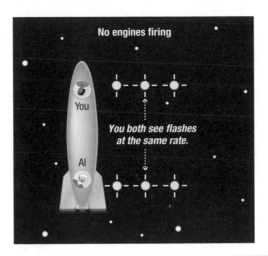

FIGURE 6.7 You and Al are floating weightlessly at opposite ends of a rocket-shaped spaceship. You both have a light that flashes once per second. Because you share the same reference frame, you'll see each other's lights flashing at the same rate. NASA/ Hubble Space Telescope Science.

We can answer this question by thinking about two facts: (1) The light flashes take a short time to travel from one end of the spaceship to the other, and (2) during that short time, the acceleration means that the speed of the spaceship increases. Let's start with your point of view. Because you are in the front of the accelerating spaceship, the increasing speed will lead you to conclude that you are being carried *away* from the point at which each of Al's flashes was emitted. Therefore, the light from each of Al's flashes will take a little longer to reach you than it did when the ship was not accelerating. As long as the spaceship keeps accelerating at the same rate, this "extra" time is always the same, which means you'll still see Al's light flashing at a steady rate—but it will now be flashing *slower* than once per second. Because you know that both lights are designed to flash once per second (and yours is still doing that), *you will conclude that Al's time at the back of the spaceship must be running slower than your time.*

Turning to Al's point of view in the back of the spaceship, the spaceship's increasing speed will mean that he is being carried *toward* the point at which each of your flashes was emitted. His viewpoint is therefore the opposite of yours. He will see each flash taking a little less time to reach him than it did when the ship was not accelerating, which means that he'll see your flashes coming at a rate of more than once per second and will conclude that your time is running *faster* than his. In other words, you and Al both agree that the lights are flashing faster at the front of the accelerating spaceship than at the back, which means you both agree that time runs faster at the front and slower at the back.

Let's pause for a moment to compare this situation to the situation we examined in chapter 5, in which you and Al were both floating weightlessly while moving past each other at high speed. In that case, you could argue endlessly about whose time was "really" running slow, because as long as your motion continued, there was no way for you to bring your clocks together to see which one had ticked off less time. In our new situation, you both share the same spaceship, which means you can easily bring your clocks together just by walking up or down the stairs. Therefore, the fact that spacetime is the same for everyone means you'll agree on what you see when you compare clocks: There will be no doubt that a clock from the back of the ship has ticked off less time than a similar clock from the front.

Now that we have found that time runs slower at the back of an accelerating spaceship, we simply apply the equivalence principle, which tells us that we would find the same results for a spaceship at rest in a gravitational field (figure 6.8). Our astonishing conclusion: For a spaceship or building or anything else on the ground, general relativity predicts that *time runs slower lower down than it does higher up*. That is, time must run more slowly at lower altitudes than at higher altitudes in a gravitational field. This effect is known as *gravitational time dilation*. The stronger the gravity—and hence the greater the curvature of spacetime—the greater the effect of gravitational time dilation.

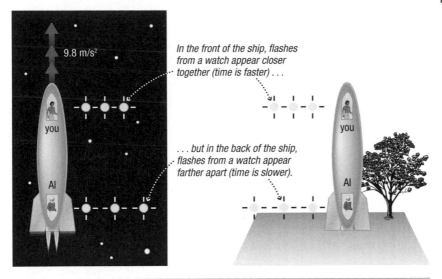

FIGURE 6.8 (*Left*) If you fire the spaceship's engines to accelerate, you effectively are moving away from Al's flashes while he is moving toward yours, which means you'll both agree that his flashes are occurring at a slower rate than yours. (*Right*) By the equivalence principle, you'd find exactly the same result for a spaceship on the ground, which means time runs slower at lower altitudes (where gravity is stronger) than it does at higher ones.

The prediction of gravitational time dilation can be tested by comparing clocks located in places with different gravitational field strengths. On Earth, precise atomic clocks have been used to measure differences in the rate at which time passes over height differences as small as a single meter. Although the differences in time that occur at different altitudes on Earth are so small that they would add up to only a few billionths of a second over a human lifetime, they agree precisely with the predictions of general relativity. On a more practical level, the global positioning system (GPS) depends on very precise comparisons between clocks on Earth and the clocks on the system's orbiting satellites. Because the satellites are moving at high speed above Earth, software that uses the system has to take into account both the effects of time dilation predicted by special relativity (due to each satellite's speed relative to the ground) and the effects of gravitational time dilation due to each satellite's altitude above the ground. These "corrections" to account for relativity are very important; without them, the positions that your GPS navigation system gives you would be noticeably inaccurate. In that sense, every time you use your navigation system, you are testing and confirming key predictions of both Einstein's special and general theories of relativity.

Given that Earth has a relatively weak gravitational field, you might wonder if we've ever tested the prediction that gravitational time dilation is greater for objects with stronger gravity. The answer is yes, because even though we've never placed clocks on any such object, nearly all astronomical objects have their own natural atomic clocks. We can observe these natural clocks by spreading light into rainbow-like spectra. With high enough resolution, we find that the spectra of the Sun and other stars contain numerous, sharp *spectral lines*. Each of these lines is produced by a particular chemical element that emits light with characteristic frequencies, which makes the lines the equivalent of atomic clocks.

To see how spectral lines allow us to test general relativity, suppose that some particular gas emits a spectral line that, when produced in a laboratory on Earth, has a frequency of 500 trillion cycles per second. If this same gas is present on the Sun, it will also emit a spectral line with a frequency of 500 trillion cycles per second. However, because the Sun has stronger gravity than Earth, general relativity predicts that time should be running more slowly on the Sun, which means that 1 second on the Sun lasts longer than 1 second on Earth. Therefore, during 1 second on Earth, we will not see all 500 trillion cycles from the gas on the Sun, which means that the spectral line will appear to have a lower frequency when we observe it in the Sun's spectrum than when we produce it in a laboratory on Earth. Because lower frequency means redder color, the slowing of time will make the lines appear redder than they would otherwise. This effect is called *gravitational redshift*, and it explains the reddening of the numerals you noticed as you dropped your clock toward your black hole in chapter 1. More important, because we know how strong gravity is on the Sun and

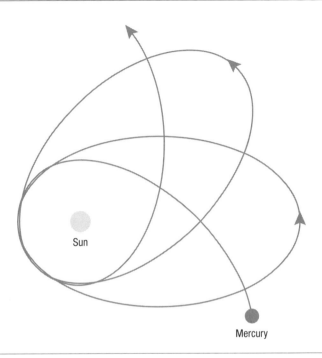

Sun

Mercury

FIGURE 6.9 This diagram shows how Mercury's elliptical orbit slowly precesses around the Sun. The picture is highly exaggerated; the actual rate of precession is less than 2 degrees per century. Newton's theory of gravity can account for most but not all of this precession; general relativity explains it fully.

other stars, general relativity allows us to predict the precise amount of gravitational redshift that we should observe. As you probably expected, the observational results match the predictions of general relativity.

General relativity has been tested in many other ways, and has so far passed every test it has been given. Here, I'll mention just one more direct test, one that had particular historical importance. Newton's law of gravity predicts that Mercury's orbit should precess slowly around the Sun because of the gravitational influences of other planets; Figure 6.9 shows a highly exaggerated view of what this precession looks like. Careful observations of Mercury's orbit during the nineteenth century showed that it does indeed precess, but calculations made with

Newton's law of gravity were not quite in agreement with the observations. The discrepancy was small (the predicted rate was off by only about 0.01 degrees per century), but astronomers could find no way to account for it. Einstein was aware of this discrepancy, and from the time he first thought of the equivalence principle, he hoped that his new idea would provide a way to explain Mercury's orbit. When he finally succeeded, he was so excited that he was unable to work for the next three days, and later called the moment of this success the high point of his scientific life. In essence, Einstein showed that the discrepancy arose because Newton's law of gravity assumes that time is absolute and space is flat. In reality, time runs more slowly and space is more curved on the part of Mercury's orbit that is nearer the Sun. The equations of general relativity take this distortion of spacetime into account, providing a predicted orbit for Mercury that precisely matches its observed orbit.

THE TWIN PARADOX REVISITED

The ideas of general relativity can be used to gain new insight into the so-called twin paradox, which we discussed briefly in chapter 4. Recall that this paradox arises when we consider twins, with one staying home on Earth while the other makes a high-speed journey to a distant star and back. Because all motion is relative, the twin who visits the star is free to claim that she goes nowhere and that it is Earth and the distant star that move, first bringing the star to her while Earth moves away, then reversing direction and bringing Earth back to her while the star moves away. The paradox surrounds the question of who ends up aging less during the trip, since each twin can claim that it was the other who traveled.

In chapter 4, we noted that the resolution of the twin paradox comes about because the twins do not have symmetrical situations: one experiences accelerations that the other does not. We found that the traveling twin ends up aging less.

With general relativity, we can use a thought experiment to look more deeply at the resolution to the paradox. Suppose you and Al are floating weightlessly next to each other and both of you have synchronized watches. While you remain weightless, Al uses his engines to accelerate a short distance away from you, decelerate to a stop a bit farther away, and then turn around and return. From your point of view, Al's motion means that you'll see his watch ticking more slowly than yours. Therefore, upon his return, you expect to find that less time has passed for Al than for you. Now let's consider how Al views the situation.

The two of you can argue endlessly about who is really moving, but one fact is obvious to both of you: During the trip, you remained weightless while Al felt weight holding him to the floor of his spaceship. Al can account for his weight in either of two ways. First, he can agree with you that he was the one who accelerated, in which case he'll agree that his watch ran more slowly than yours because time runs more slowly in an accelerating spaceship. Alternatively, he can claim that he felt weight because his engines counteracted a gravitational field in which he was stationary while you fell freely. Note, however, that he'll still agree that his watch ran more slowly than yours, because time also runs slowly in gravitational fields. No matter how either of you looks at it, the result is the same: Less time passes for Al.

The left side of figure 6.10 shows a spacetime diagram for this experiment. You and Al both moved between the same two events in spacetime (the start and end points of Al's trip). However, your path between the two events is shorter than Al's. Because we have already concluded that less time passes for Al, we are led to a remarkable insight about the passage of time: *Between any two events in spacetime, more time passes on the shorter (and hence straighter) path.* The maximum amount of time you can record between two events in spacetime occurs if you follow the straightest possible path—that is, the path on which you are weightless.

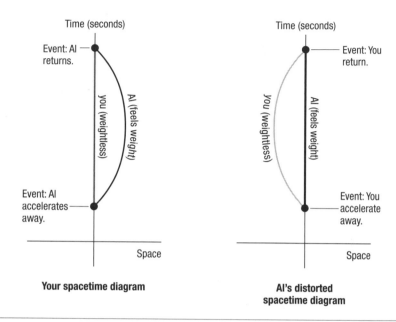

FIGURE 6.10 You and Al move apart and come back together, with you floating weightlessly the whole time while Al feels weight. Because you are weightless, you can draw your spacetime diagram as shown on the left, concluding that more time passes for you because you follow the shorter and straighter path. Al explains his weight as being due to gravity, which means he would have to draw his spacetime diagram on a curved piece of paper. Therefore, if we show it flat, as on the right, it must be distorted; if we viewed it with the correct curvature, Al's diagram would also show that you follow the shorter and straighter path.

The twin paradox is really nothing more than the question of why Al can't claim to be the one who follows the shorter and straighter path. After all, he would surely be tempted to draw the spacetime diagram on the right in figure 6.10, in which he draws his own world-line going straight up through time because he considers himself to be at rest. His diagram makes it *appear* that he has the shorter and straighter path, but this appearance is actually a distortion of reality. Remember that the only way he can claim to be at rest is if he asserts that his weight is due to gravity, and in that case the gravity must curve spacetime in his vicinity. Therefore, if he wants to draw a spacetime diagram that shows himself "at rest," he would have to do it on an appropriately curved piece of paper—one that would show that his path is indeed longer and more curved than yours.

Al's problem is analogous to that of a pilot who plans a trip from Philadelphia to Beijing with a flat map of Earth. On the flat map, the straightest possible path *appears* to go along the line of latitude that connects the two cities, as shown by the straight path in figure 6.11. However, this map is distorted, because Earth's surface is actually curved. The shortest and straightest path is still the great-circle route that we saw in figure 6.2, even though this path appears curved and longer on the flat map. Just as the distortions in a map of the world do not change the actual distances between cities, the way we choose to draw a spacetime diagram does not alter the reality of spacetime. Al really is the one for whom less time passes, because his path through spacetime really is longer and more curved.

To be sure these ideas are clear, let's revisit your trip to the black hole. In order to get up to speed—and to turn around and come

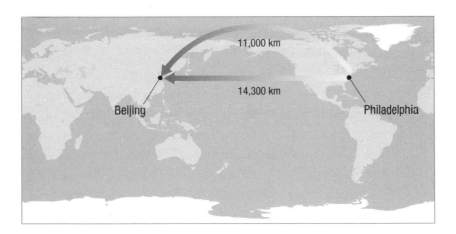

11,000 km

14,300 km

Beijing

Philadelphia

FIGURE 6.11 This flat map shows the same two paths between Beijing and Philadelphia that we saw in figure 6.2. The distortion involved in making a flat map of the round Earth makes the actual shortest and straightest path look like it is longer.

home—you underwent tremendous accelerations. These accelerations are the equivalent of extremely strong gravity, and during the periods of acceleration general relativity tells us that your clocks must have ticked off much less time than clocks back on Earth. That is why you came home having aged less than people on Earth. There's some good news here. Recall that we set up your trip so that your accelerations were nearly instantaneous. This had the advantage of making it easy to calculate how much time dilation would occur for you as seen by people on Earth (because we could use the simple formula from special relativity), but it had the disadvantage of meaning that the forces of acceleration would have killed you. We can now offer you a safer trip: Instead of instantaneous acceleration, you could accelerate gradually until the midway point, then decelerate gradually to the black hole, and then do the reverse on the return trip. As long as your *average* speed is the same as we assumed before, then the overall time for the trip will still unfold just as we found previously.

GRAVITATIONAL WAVES

We have just one major mystery left to address with general relativity, and it is the mystery of how events that occur in one place affect objects in other places. More specifically, we know that masses cause the curvature of space and spacetime, and that masses are always in motion, whether through orbits, or explosions, or otherwise. Just as masses moving in circles on a rubber sheet will affect the precise shape of the sheet even at great distances from the masses, changes in the local curvature of space in one place must ultimately affect the curvature in other places. But how is the fact that a change has occurred in one place communicated out to other places?

When Einstein analyzed this question, he found that changes to the curvature of space in one place propagate outward to other places like ripples on a pond. For example, the effect of a star suddenly imploding or exploding should be rather like the effect of dropping a rock into a pond, and two massive stars orbiting each other closely and rapidly should generate ripples of curvature in space rather like those of a blade turning in water. Einstein called these ripples *gravitational waves*.

Gravitational waves are predicted to be similar in character to light waves in that they have no mass and should travel at the speed of light. In much the same way that light waves cause charged particles (such as electrons) to move back and forth as they go by, the distortions of space carried by gravitational waves should compress and expand any masses that they pass by. We should in principle be able to detect these waves by looking for this type of compression and expansion, but there's a problem: Gravitational waves are expected to carry far less energy than light waves, which means we would need incredibly precise measurements to detect their effects on objects on Earth. As of 2013, no one has succeeded in making an unambiguous detection of gravitational waves, though several major efforts are under way. The best known is called the Laser Interferometer Gravitational-Wave Observatory (LIGO), which currently consists of large detectors in Louisiana and Washington State that search in tandem for telltale signs of gravitational waves. NASA and the European Space Agency have also developed preliminary plans for a much more sensitive space-based gravitational wave observatory called the Laser Interferometer Space Antenna (LISA). However, due to budget constraints, it is unlikely that this observatory could be launched before about 2025 at the earliest.

Given that gravitational waves are an important prediction of the general theory of relativity, should we be concerned that they have not yet been observed? Most scientists think not, because while we lack a direct detection of gravitational waves, we have very strong indirect evidence of their existence. This evidence comes from "binary pulsars," which are much like other binary star systems (meaning two stars orbiting each other) except that both stars are highly compressed *neutron stars*.

Neutron stars are incredibly dense; they are typically more massive than our Sun but only about 20 kilometers in diameter. (For comparison, our Sun's diameter is about 1.4 million kilometers). Their small sizes allow them to orbit each other much more closely and rapidly than ordinary stars, and general relativity predicts that such systems should emit significant amounts of energy in the form of gravitational waves. This emission of gravitational waves means the system as a whole gradually loses energy, and this energy loss should cause the orbits of the two neutron stars to decay.

The first known binary pulsar was discovered by Russell Hulse and Joseph Taylor in 1974. Hulse and Taylor carefully monitored the orbits of the two neutron stars and found that the orbital period is indeed decreasing as though the system is losing energy. Moreover, the orbital period is decreasing at precisely the rate predicted when we assume that the energy loss is due to gravitational waves. These observations so strongly suggest that gravitational waves really exist that Hulse and Taylor were awarded the Nobel Prize in physics in 1993. Continued observations of the Hulse-Taylor binary have only further confirmed the prediction of general relativity, and astronomers have since found similar systems that are providing additional verification.

WAS NEWTON WRONG?

We have touched on several major cases in which Einstein's general theory of relativity makes predictions that differ from the predictions of Newton's older theory of gravity. In every one of these cases, observations indicate that Einstein's theory is the one that is correct. Given this fact, it's worth asking: Does the success of Einstein's theory mean that Newton's theory was wrong?

To some extent, the answer depends on how you define "wrong," but the question gives important insights into the nature of science. For all those cases in which Einstein's theory and Newton's theory give different answers, the observations clearly show that Newton's answers are incorrect. But it's important to remember that for most cases, the answers given by the two theories are nearly indistinguishable. That is why astronomers still use Newton's law of gravity to calculate the orbits of planets around stars, of stars around the centers of their galaxies, and of galaxies around each other. It is also why, during the famous mishap with the *Apollo 13* mission (chronicled in the movie of the same name), astronaut Jim Lovell could correctly say, "We just put Sir Isaac Newton in the driver's seat" when they shut down all their engines. For the vast majority of the cases that we ever encounter, Newton's theory of gravity works very nearly as well as Einstein's.

From that standpoint, the issue of right or wrong for a theory comes down to scientific demands for testability. When the term "theory" is used properly, it refers to an idea that has been very well tested and verified. But just because a theory has passed all the tests given to date doesn't necessarily mean it will continue to pass future tests. Consider the old adage "What goes up must come down." If we think of this as part of a theory of motion on Earth, it's a pretty good one; try as you might, you'll never be able to throw something hard enough that it does not come down. Nevertheless, once Newton developed his theory of gravity, we learned that the old theory was incomplete: It works fine for ordinary tossed objects, but breaks down if you can launch an object with enough speed so that it reaches escape velocity. Equally important, Newton's theory of gravity gave us a new mental picture of objects going up and down, one in which motion under gravity was no longer restricted to tossed objects on Earth but was instead extended to the motions of objects in the heavens.

In the same way, Einstein's theory doesn't invalidate any of the many cases in which Newton's theory works just fine; it only shows us that there are some cases in which Newton's theory is not good enough, while giving us a new mental picture of gravity that eliminates the "action at a distance" idea that Newton himself regarded as absurd. Einstein's theory may also be incomplete; indeed, as we'll discuss in chapter 7, it seems to break down when we try to apply it to the center of a black hole. For scientists, such breakdowns are where the excitement lies, because they can point us to new theories that may provide even deeper insights into nature. But if and when we find a better theory, the large foundation of evidence supporting Einstein's theory will remain standing. This places a heavy constraint on any replacement theory, since it must give the same answers as Einstein's theory for all the cases in which general relativity works.

So to summarize, my own answer to the question "Was Newton wrong?" is *no*. Newton gave us a powerful theory of gravity and motion, and he was as "right" as he could have been given the observations and experiments possible in his time. Science is a great edifice, built one brick at a time. As long as we lay the bricks carefully, we will always be able to build further, and we will not need to remove the bricks already placed. For a somewhat more elegant metaphor, we can turn to the words of Sir Isaac Newton himself: "If I have seen farther than others, it is because I have stood on the shoulders of giants." Through his theories of relativity, Einstein joined Newton and the other giants, and someday others will stand upon his shoulders as well.

Part 4

IMPLICATIONS OF RELATIVITY

7

BLACK HOLES

OVER THE past several chapters, we have discussed the reasons behind many of the phenomena you experienced during your voyage to a black hole. We have seen why less time passed for you during the journey than passed for people back home on Earth. We've learned that the structure of spacetime at a distance from any mass depends only on the amount of mass, which is why you can orbit a black hole at a distance just as you can orbit a star, without fear of being sucked in. We've found that the slowing of time and the gravitational redshift you observed when watching your clock fall toward the black hole are expected consequences of the simple ideas that lie at the heart of Einstein's theories, including the ideas that all motion is relative, that the speed of light is absolute, and that effects of acceleration are equivalent to effects of gravity.

What we have not yet addressed is the question of exactly what a black hole *is*, and what your colleague encounters as he plunges across the event horizon. So let's turn to these questions now, and complete the story of the journey that you embarked on in chapter 1.

HOLES IN THE UNIVERSE

Look back at figure 6.3, which shows a rubber sheet analogy explaining the orbits of planets around the Sun, and imagine what would happen if the Sun somehow became more compressed, without changing its total mass. You can get the idea by picturing what would happen if you placed a denser weight on the rubber sheet; for example, replacing a 5-kilogram bowling ball with a 5-kilogram iron shot. Clearly, the rubber sheet would become much more deformed at the location of the denser weight. However, if you look to regions relatively far from the weight, the deformity of the rubber sheet would be unchanged, because the total weight was unchanged.

Figure 7.1 shows the idea. The left diagram shows the Sun on a rubber sheet. The middle diagram shows how the rubber sheet changes if the Sun becomes more compressed. If you continue to compress the Sun, it will press farther and farther down on the rubber sheet, which in our analogy corresponds to more and more curvature of spacetime. If you compressed the Sun enough, it would eventually push down on

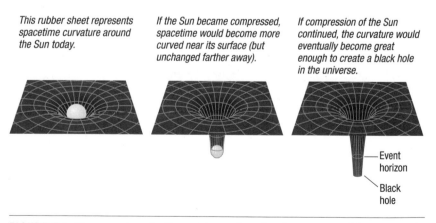

This rubber sheet represents spacetime curvature around the Sun today.

If the Sun became compressed, spacetime would become more curved near its surface (but unchanged farther away).

If compression of the Sun continued, the curvature would eventually become great enough to create a black hole in the universe.

Event horizon

Black hole

FIGURE 7.1 Compressing the Sun without changing its mass would increasingly deform the region of spacetime near the Sun. If the Sun became dense enough, it would essentially create a hole in the observable universe—a black hole.

the rubber sheet until it opened up a hole in the sheet. Although the analogy breaks down at this point (and again, you shouldn't take any of the rubber sheet analogies too literally), the general idea still applies to spacetime: With sufficient compression of the Sun, spacetime would become curved so much that it would in essence create a hole in the observable universe. The name *black hole* should now make sense: It is *black* because not even light can escape from it, and it is a *hole* in the sense that objects falling into it can no longer be observed by us with any conceivable technology.

THE EVENT HORIZON

Black holes have an inside and an outside, as you can understand by thinking back to the clock with a rocket strapped to it that you dropped toward the black hole in chapter 1. When you first sent it out of your spaceship, it would have been relatively easy for the rocket to stop the clock's fall and reverse it. As it fell closer to the black hole, the stronger gravity—or, equivalently, the greater curvature of spacetime—would have required the rocket to fire much more strongly to stop and reverse the fall. Eventually, the clock and rocket would reach a "point of no return" at which no amount of force could prevent their continuing fall, and from which not even light can escape to return to the outside universe. This point of no return is the *event horizon* that we first discussed in chapter 1;[1] it gets its name because events that occur within it can neither be seen from the outside universe nor influence the outside universe.

Note that you will often hear the event horizon described as the place where gravity becomes so strong that the escape velocity reaches the speed of light. However, this is not a great way to think about it,

---

1. Here, I am assuming a nonrotating black hole; rotating black holes have a more complex spacetime geometry that introduces some additional effects near the event horizon (which actually splits into an inner and outer horizon), but they do not alter the basic ideas in our discussion.

because it makes it sound like light beams would "almost" be able to escape in much the same way that a rocket with a speed short of the escape velocity "almost" escapes Earth. But unlike the rocket, which starts out going up, then slows and returns, light must always travel at the speed of light. As a result, a much better analogy is a river with a waterfall, in which space itself flows toward the black hole.[2] (The idea of "flowing space" may sound strange, but it turns out to be a mathematically accurate description of the behavior of spacetime near a black hole.) Far from the black hole, the "river" of space flows so slowly that you don't even notice it. But as you approach the event horizon, the river moves ever faster, making it more and more difficult to paddle against the current. The event horizon itself is the waterfall, at which space "goes over the cliff" in the sense of flowing into the black hole so fast that even a paddler traveling at the speed of light would still be carried over the cliff. This analogy brings us back to our earlier discussions of black holes and "sucking." Far from the black hole, where there is no noticeable flow of space, orbits can be calculated from Newton's universal law of gravitation, and it's very clear that the black hole will not "suck you in." As you get very close, however, the flow of space will eventually become so strong that it may feel as though you *are* being "sucked in." Nevertheless, the black hole is no more sucking you in than is the pool at the bottom of a waterfall. You go over a waterfall because that is the way the river around you is flowing, and you fall into a black hole because you are carried along with the space around you; there's no cosmic vacuum cleaner that is sucking you in.

Another way to think about the event horizon is by applying what we've learned about time dilation and gravitational redshift in general relativity. Recall that the stronger the gravity, the more time slows and

---

2. This analogy, developed by University of Colorado professor Andrew Hamilton, is beautifully depicted in the planetarium show and movie *Black Holes: The Other Side of Infinity* (for which Hamilton was the science director). It is also depicted on Hamilton's "Inside Black Holes" website.

the more light becomes redshifted. If you take this idea to its limit, you might imagine that there could be a place where gravity would be so strong that, at least from the perspective of an outside observer, time would come to a stop and light would become infinitely redshifted. Strange as this idea sounds, it describes what we would observe at the event horizon of a black hole. That is why, as your clock fell toward the event horizon of the black hole in chapter 1, its light became more and more redshifted until it disappeared from view, and as it did so, you realized that time was coming to a stop on the clock face.

We now come to the last idea from chapter 1 that we have not yet explained: the fate of your overeager colleague who made a dive for the black hole. If we ignore his death due to tidal forces, from his own point of view he plunges across the event horizon and into the black hole in little time at all. This should now make sense. Recall that when your rocket-shaped spaceship was accelerating through space or on the ground in gravity (see figure 6.8), you observed Al's time to be running slow, but he considered his own time to be running normally. This same idea just becomes more extreme as your colleague plunges toward the black hole. From your point of view in orbit, his time runs slower and slower until it comes to a stop at the event horizon, which is why you would never actually see him reach or cross it. From his point of view, however, his own time always seems to run normally, and he won't feel anything special as he races across the event horizon. He'll continue to plunge rapidly toward the center of the black hole.

To summarize, the event horizon in essence marks the boundary between the inside of a black hole and the outside universe. Viewed from the outside, the event horizon has three crucial properties: It is the place where it becomes impossible to return to the outside universe, it is the place where time appears to come to a stop, and it is the place where light becomes infinitely redshifted. But it is not a physical boundary. For an object falling into a black hole, the event horizon is merely the place beyond which it can no longer contact the outside universe as it proceeds toward whatever fate awaits it within the black hole.

BLACK HOLE PROPERTIES

As we've discussed, the Sun in principle could become a black hole if we compressed it enough. What, then, would have become of the material that made up the Sun? It would have disappeared into the black hole, and therefore would no longer be "material" in any ordinary sense. In essence, the former Sun would have become a disembodied mass causing curvature of spacetime.

This leads to the question of what you would see if you looked at a black hole. Don't let our rubber sheet diagrams fool you—the funnel-shaped hole is only a two-dimensional analogy. In reality, if you were close enough to a black hole to be able to see it, you would see a three-dimensional black sphere with a size defined by its event horizon, which would also be spherical.[3] In principle, you could measure the circumference of the event horizon, and from that you could calculate the radius of a circle with that circumference. This radius, called the *Schwarzschild radius*, is what we usually use to describe the size of a black hole. It gets its name from the fact that it was first calculated by Karl Schwarzschild, who made the calculation within a month of Einstein's publication of the general theory of relativity. Sadly, Schwarzschild died less than a year later, from an illness contracted in his role as a German soldier in World War I.

The Schwarzschild radius of a black hole depends only on its mass, and the formula for calculating it turns out to be very easy to use: It is approximately 3 kilometers times the black hole's mass in solar masses. For example, a 1-solar-mass black hole (that is, a black hole with the same mass as our Sun) has a Schwarzschild radius of about 3 kilometers, a 10-solar-mass black hole has a Schwarzschild radius of about 30 kilometers, and a 1-billion-solar-mass black hole has a Schwarzschild radius of about 3 billion kilometers. Keep in mind that while the

---

3. The black hole would be a perfect sphere only if it was not rotating; a rotating black hole would be stretched into an ellipsoidal (football-like) shape. Also note that although the black hole's actual shape is simple, the light patterns that you would see around a black hole would be quite complex due to the fact that its gravity strongly bends light that passes near it.

Schwarzschild radius can be used to describe the circumference of the event horizon or the volume of space that will appear to be occupied by the black hole, you could not actually measure a Schwarzschild radius directly. The reason, of course, is that spacetime is so distorted inside the event horizon that the idea of a radius no longer has meaning.

The idea that a black hole is essentially a disembodied mass means that black holes are very simple objects, at least in terms of what we can learn about them from the outside. For example, imagine that you had two objects with the mass of the Sun, one an ordinary star and the other a giant diamond. If both somehow collapsed to form black holes, you'd no longer be able to tell any difference between them; both would simply be black holes with masses equal to the mass of the Sun.

In fact, besides its mass, a black hole can retain only two other properties of whatever made it or has fallen into it: electrical charge and rotation rate. Electrical charge is not expected to play much of a role, because any net positive or negative charge that a black hole might have would be neutralized quickly as it attracted oppositely charged particles from the surrounding interstellar gas. Rotation has significant effects near the event horizon, and because we expect most black holes to be rotating rapidly (as a result of the way in which they are thought to form), physicists who study black holes must take these effects into account. However, rotation has very little effect once you move away from the immediate vicinity of the event horizon, and we will not discuss its effects in this book.

TOO STRANGE TO BE TRUE

Although Schwarzschild discovered the formula for his famous radius in 1916, most astronomers doubted that black holes could really exist for many decades after that.[4] The main problem was that the idea of a

---

4. The term "black hole" was not actually used until 1967, when it was coined by John Archibald Wheeler. Before that, scientists used a variety of other terms (such as "collapsar" or "dark star") to describe the same idea.

black hole just seemed too strange to be true. But a secondary problem was that while it was easy to calculate Schwarzschild radii, astronomers did not know of a way that a real object could ever become so highly compressed.

To understand this idea, we need to look a little deeper into the meaning of the Schwarzschild radius. Although nowadays we usually associate it with the size of actual (or suspected) black holes, it's really just a number that tells us how small we'd need to compress a mass to turn it into a black hole. For example, when we say that the Sun has a Schwarzschild radius of 3 kilometers, what we really mean is that the Sun would become a black hole if we could somehow squeeze it down until its current radius of 700,000 kilometers became a mere 3 kilometers; at that point, the Sun would disappear into its own event horizon. In fact, you can calculate a Schwarzschild radius for *any* mass. Earth, with a mass about 1/300,000 that of the Sun, has a Schwarzschild radius of about 1 centimeter (which is 1/300,000 of 3 kilometers), meaning that Earth would become a black hole if we could compress it to the size of a marble. Even *you* have a Schwarzschild radius, which turns out to be about 10 billion times smaller than an atomic nucleus; in other words, if you were somehow compressed to that tiny size, you would disappear from the universe into your own mini black hole.

The key question for the existence of black holes, then, is whether there is any way for nature to compress objects down to sizes smaller than their own Schwarzschild radii. An important clue that this might in some cases be possible came with calculations performed in 1931 by physicist Subrahmanyan Chandrasekhar (for whom NASA's Chandra X-ray observatory was named). By that time, astronomers had recognized the existence of numerous *white dwarf* stars, which are objects with masses similar to that of the Sun but compressed to sizes no larger than Earth. The high density of white dwarfs—a typical teaspoon of white dwarf material would outweigh a small truck if you brought it to Earth—was already surprising to astronomers, but Chandrasekhar's calculations suggested that much denser objects might also exist. In particular, he found that there was a *maximum* possible mass for a white dwarf; later refinements to his original calculation showed that this *white dwarf limit* (also called the *Chandrasekhar limit*) is about 1.4 times the mass of the Sun. This implied that if a white dwarf had a mass greater than this limit, it could no longer support itself against its own gravity and it would therefore collapse much further.

Within a few years after Chandrasekhar published his work, several other scientists independently worked out what would happen to a white dwarf that grew in mass until it exceeded the white dwarf limit. They found that such an object would then collapse until the electrons and protons that made up its atoms combined to form a ball of neutrons, creating what we call a *neutron star*. Most astronomers considered neutron stars too strange to be true, but at least a couple (notably Fritz Zwicky and Walter Baade) suggested that they might be a natural by-product of supernova explosions—the titanic explosions that occur at the ends of the lives of massive stars—an idea that later proved to be correct. (Recall from chapter 6 that we've even discovered binary systems consisting of two neutron stars, and that the orbital decay of these systems provides strong evidence for the existence of gravitational waves.)

In case you're wondering why the idea of a neutron star seemed so strange, note that a typical neutron star has a mass greater than that of the Sun packed into a volume only about 10 kilometers in radius. You can use this fact to show that it has an incredible density—a teaspoon of neutron star material would outweigh a large mountain on Earth—but an even better way to get a sense of a neutron star's awesome gravity is to think about what would happen if one magically appeared on Earth. Because of its relatively small *size*, a neutron star would fit easily within the boundaries of many large cities. But it would not just sit there. Instead, because a neutron star's mass is hundreds of thousands of times greater than that of the entire Earth, Earth would "fall" onto the surface of the neutron star, becoming compressed to neutron star density in the process. By the time the dust settled, the former Earth would have been squashed into a spherical shell no thicker than your thumb on the neutron star's surface.

To return to our main story, the fact that few scientists thought neutron stars could be real didn't stop them from trying to calculate their properties. In 1938, Robert Oppenheimer, who later became the leader of the Manhattan Project (the World War II project that built the first atomic bomb), decided to investigate whether neutron stars might have a maximum mass of their own. He and his colleagues soon concluded that the answer was yes. They found that if a neutron star had a mass exceeding just a few solar masses, not even the internal pressure exerted by its neutrons could stop the crush of gravity. Because no known force could supply a greater pressure, Oppenheimer speculated that gravity would then crush the matter into a black hole.

As always in science, the question of whether neutron stars or black holes really exist would have to be settled by evidence. The first key evidence came from the study of white dwarfs. Over the next few decades, astronomers discovered many more white dwarfs. Not a single one had a mass exceeding Chandrasekhar's calculated limit, suggesting that his limit was real. Because many stars are more massive than the 1.4-solar-mass limit, the idea that some stars might ultimately collapse into neutron stars began to be taken more seriously.

A pivotal moment came in 1967, when a British graduate student named Jocelyn Bell discovered the first *pulsar*—an astronomical object emitting radio waves that pulsate with astonishing regularity. Her first pulsar was blinking with radio waves every 1.3 seconds. The timing of the blinks was more precise than that of any human-built clock at the time, and before it was called a pulsar some astronomers only half-jokingly called it an "LGM" for "little green men." Within about a year, however, astronomers had figured out what was really going on. Further searches turned up pulsars sitting in the centers of supernova remnants, which are the remains of stars that exploded in supernovas. Putting two and two together, astronomers realized that pulsars were rapidly rotating neutron stars. The reason for the pulsations is that neutron stars tend to have strong magnetic fields that lead them to beam radiation along their magnetic axis. Therefore, if the magnetic axis is tilted relative to the rotation axis (as is the case for Earth, in which the magnetic poles are offset from the geographic poles by many hundreds of kilometers), the beams sweep around with each rotation, much like the rotating beam from a lighthouse. If the tilted axis happens to be oriented so that one of the beams sweeps by Earth, then we see a pulse of radiation with each rotation. The rapid rotation rate also serves to confirm the small size and incredible density of neutron stars. We know that the objects cannot be much larger than the presumed size of neutron stars, because a larger radius with such a fast rotation rate would imply a surface moving faster than the speed of light. We know the objects must be as dense as we expect for neutron stars, because if they were less dense, their gravity would be too weak to prevent them from breaking apart at their fast rotation speeds.

Confronted with clear evidence that "too strange to be true" neutron stars really exist, astronomers became more open-minded about the possibility that black holes might also exist. It wasn't long before the evidence began to indicate that this is indeed the case.

ORIGINS OF BLACK HOLES

If you want to find evidence for the existence of black holes, there are really two steps. First, you need to find objects that are incredibly dense but that exceed the limiting mass of neutron stars. Second, you need to convince yourself that such an object would indeed be a black hole, rather than some other weird state of matter that is simply more highly compressed than a neutron star. The first step can be done through observations, but the second requires an understanding of how black holes might actually form.

The key to understanding black hole formation comes in recognizing that *all* astronomical objects are engaged in a perpetual struggle between the force of their own gravity, which tries to make them smaller, and internal forces that generate pressure to push back against the crush of gravity.

Let's start with Earth. Gravity holds Earth together and shapes our planet into a sphere. If you think more deeply about it, however, you might wonder why gravity stopped where it did. That is, why didn't gravity compress our planet to a higher density, or for that matter, keep compressing it until it became a black hole? The answer is that Earth is made of atoms, and the forces between these atoms (which arise from electromagnetic forces acting between the charged particles that make up atoms) grow stronger as you try to squeeze them closer together. Earth is the size that it is because this is the size at which Earth's gravity comes into natural balance with the interatomic forces resisting gravity.

The same idea holds for other planets, as well as for moons, asteroids, and comets. Their sizes are always determined by the balance between the inward force of gravity and the outward force of the pressure generated by the fact that atoms tend to resist being squeezed together. The results can sometimes be surprising, especially when we consider planets made mostly from hydrogen and helium, which tend to compress much more readily than rock and metal. For example, Jupiter has more than three times the mass of Saturn, but the two planets are nearly the same size. The reason is that if you took Saturn and started adding more hydrogen and helium to it, the increasing strength of gravity would compress it to higher density; it would therefore grow in mass with very little change in its size.

This brings us to our next "deeper question." Stars are nearly identical in composition to planets like Jupiter and Saturn, being made almost entirely of hydrogen and helium. What, then, makes Jupiter a planet and the Sun a star? To answer this question, let's think about what would happen if we took a planet like Jupiter and kept adding mass to it. The more mass we added, the stronger its gravity would become, and this stronger gravity would gradually compress the object's central core to both higher density and higher temperature. Eventually, the core would become so hot and dense that hydrogen nuclei would crash into each other hard enough to fuse together. This is the process of nuclear fusion that makes stars shine. Recall that hydrogen fusion turns hydrogen into helium and generates energy in accord with $E = mc^2$, because helium nuclei are slightly less massive than the hydrogen nuclei that make them. Summarizing, all the differences between a Jupiter-like planet and a star are traceable solely to mass. With enough mass,[5] any ball of hydrogen and helium gas will inevitably become a star.

The size of a star is determined by the same type of balance between gravity and internal pressure that determines the size of planets. In the case of a star, however, most of this pressure arises from the flow of the energy generated by nuclear fusion. That is, the energy generated by fusion keeps the gas particles within a star moving at high speeds, and the constant collisions between these particles create pressure that supports the star against the crush of gravity. (Additional pressure is supplied by photons of light carrying energy within the star; this *radiation pressure* plays a particularly important role in high-mass stars.)

The basic problem that a star must face is that fusion cannot continue to generate energy forever. As a star's life progresses, it gradually converts more and more of its core hydrogen into helium, which means the hydrogen will eventually run out. The time until this occurs depends on the mass of the star. Although it may seem counterintuitive, high-mass stars live *shorter* lives than lower-mass stars. The reason is that the rate at which fusion occurs in a star's core is very sensitive to temperature, with relatively small temperature increases leading to large increases in the fusion rate. The added crush of gravity in higher mass stars makes their cores hotter and their fusion rates much higher, which explains why massive stars shine far more brightly than their lower-mass cousins. In fact, high-mass stars burn through their hydrogen at such a prodigious rate that they can run out of it in just a few million years. In contrast, a lower-mass star like our Sun can shine steadily for 10 billion years before its core hydrogen runs out, and stars with masses lower than that of the Sun live even longer.

---

5. The minimum mass needed to make a star is about 8% of the Sun's mass, which is equivalent to about 80 times the mass of Jupiter. If the mass is less than this, then gravity is not strong enough to compress the core to the temperatures and densities needed for sustained nuclear fusion.

Regardless of when it happens, the end of hydrogen fusion means the end of the pressure that supported the star's core against the crush of gravity. The core, now composed primarily of helium, must therefore begin to contract.[6] This raises the core's temperature and density further, and at some point it will become so hot and dense that its helium nuclei will begin to fuse. The basic helium fusion process fuses three helium-4 nuclei into one carbon-12 nucleus, so the core now gradually transforms from one made mostly of helium into one made mostly of carbon. (The numbers after the element names are atomic masses, which are the sum of the numbers of protons and neutrons in each nucleus; helium-4 nuclei consist of 2 protons and 2 neutrons, while carbon-12 nuclei consist of 6 protons and 6 neutrons.) Like hydrogen fusion, the helium fusion process generates energy; it therefore gives the star a new source of internal pressure, and this pressure halts the gravitational contraction. But the reprieve is only temporary, because the helium must also run out. When it does, the relentless crush of gravity will once again cause the core to begin contracting. Typically, a star can fuse helium for about 10% as long as it fused hydrogen.

What happens next depends on the star's mass. For a relatively low-mass star like our Sun, the mostly carbon core is basically the end of the line. Before the core ever gets hot enough to fuse carbon, its contraction will be halted by a form of pressure very different from the pressure that supports the star when it is generating energy through fusion. This form of pressure is called *electron degeneracy pressure*, and it becomes the dominant source of pressure when the core attains the density of a white dwarf.

---

6. Readers who have studied astronomy will know that while the core is contracting, the outer layers of the star actually begin to expand, ultimately turning the star into a *red giant*. The reason for this expansion is that although the hydrogen has run out in the central core, which is now made mostly of helium, plenty of hydrogen still remains atop the central core. The contraction of the core and the surrounding layers raises the temperature enough that hydrogen fusion can begin in a layer surrounding the helium region, and this fusion actually proceeds at such a high rate (due to the high temperature) that it causes the star's outer layers to expand.

The precise nature of electron degeneracy pressure is somewhat peripheral to our story about black holes, so I won't spend much time on it in this book. However, for those who know something about chemistry, electron degeneracy pressure is a form of pressure that arises for essentially the same reason that two electrons cannot share precisely the same energy level in an atom (which is described in technical terms by the *exclusion principle*). In other words, the same property of electrons that explains the arrangement of the elements in the periodic table that you learn about in chemistry class also explains the pressure that arises when electrons resist being forced too closely together in a collapsing stellar core.

Electron degeneracy pressure explains the fate of low-mass stars. At essentially the same time that this pressure stops the core contraction, the star also sheds its outer layers into space. For a few thousand years, these outer layers will be visible as an expanding shell of gas around the star; we call these shells *planetary nebulae* (figure 7.2), although they have nothing to do with planets. This shedding leaves the stellar core exposed, and because electron degeneracy pressure has stopped the core collapse at white dwarf density, the core essentially *is* a white dwarf at this point. In other words, white dwarfs are the "dead" remains of low-mass stars, and they will not collapse any further because they are permanently supported by electron degeneracy pressure. We can also now explain why white dwarfs never have masses above the 1.4-solar-mass limit first calculated by Chandrasekhar. Remember that the more massive the star, the stronger the gravitational force trying to crush the core. Chandrasekhar's calculations showed that if the core's mass exceeds 1.4 times the mass of our Sun, its own self-gravity would be so strong that electron degeneracy pressure could no longer hold off the collapse.

In a sense, white dwarfs represent a permanent truce between pressure and gravity. But because this truce can hold only in relatively low-mass stars, we now need to turn to the question of what happens in higher-mass stars.

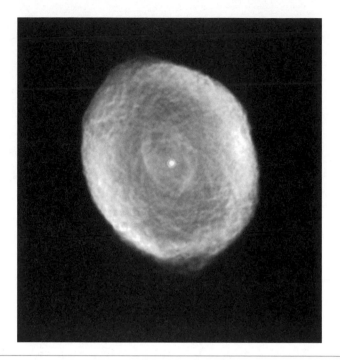

FIGURE 7.2 This Hubble Space Telescope image shows the Spirograph Nebula, an example of a *planetary nebula* created by the expanding shell of gas ejected from a dying, low-mass star. The remaining stellar core, visible in the center of the nebula, is a white dwarf, an object in which the crush of gravity is held off by electron degeneracy pressure. NASA/ Hubble Space Telescope Science Institute.

For more-massive stars, the density and temperature of the collapsing core *do* eventually become great enough to trigger the start of carbon fusion, providing the star with another temporary reprieve from the crush of gravity. A major product of carbon fusion is oxygen (because the process usually entails the fusion of helium-4 with carbon-12, producing oxygen–16). As the carbon runs out, the oxygen can begin to fuse, mostly into neon, and then the neon can begin to fuse, and so on. However, in a cosmic analog to the famous tag line of *Star Trek*'s alien race called the Borg, gravity might as well be saying, "Resistance is futile." Each new round of fusion lasts a shorter time than the next, and one day the product of fusion will be the element iron. This day will be the star's last.

The problem is that the star cannot generate energy by fusing iron. In fact, fusion of iron actually saps energy, rather than generating it. Therefore, at the instant the iron core begins to collapse, it catastrophically implodes. In a fraction of a second, this implosion releases so much gravitational potential energy that it causes the rest of the star to *explode*, creating the explosion that we call a *supernova*. Supernovae are spectacular events that you can read more about in any astronomy book. Here, our focus is on what happens to the imploding core.

We already know that the core is too massive to have its collapse halted by electron degeneracy pressure, and because electron degeneracy pressure is the final line of resistance for electrons being squeezed together, the electrons are essentially left with no choice but to combine with protons to make neutrons. The imploding core essentially becomes a ball of neutrons, which is what we've already defined as a neutron star. In other words, neutron stars are formed by supernovae, which explains why we often observe neutron stars in supernova remnants. But that is not the only possible outcome of a supernova.

The pressure that halts the crush of gravity in a neutron star is called *neutron degeneracy pressure*, and it is analogous to electron degeneracy pressure except that it occurs with neutrons. As we discussed earlier, Oppenheimer and his colleagues found that there's also a limit to how much gravity can be fought off by neutron degeneracy pressure, and modern calculations show that this limit occurs when the core's mass is around 3 solar masses. So what happens if the imploding core exceeds this neutron star limit?

In that case, the pressure exerted by the neutrons will still be short of what is needed to halt the crush of gravity, which means the core must continue its implosion. If you think black holes are too strange to be true, you might then hope that some other strange state of matter would form and create another type of pressure to stop the collapse. But no such matter or source of pressure is known. Moreover, there are two major reasons why it is unlikely that any such matter or pressure could exist.

The first reason is a simple size argument. Remember that a typical neutron star has a radius of only about 10 kilometers, while the Schwarzschild radius of a 3-solar-mass black hole is about 9 kilometers (which is 3 times the 3-kilometer Schwarzschild radius for a 1-solar-mass black hole). This means there just isn't much room in between a neutron star and a black hole. That is, if a neutron star collapses only a little more, it would disappear within its own event horizon and become a black hole.

The second and more compelling reason is that gravity has one last trick up its sleeve. Recall that what $E = mc^2$ really tells us is that mass and energy are ultimately equivalent. Therefore, energy must in principle be a source of gravity just like mass. For most objects, the gravity of their internal energy is negligible, but that is not the case in the extreme conditions of an imploding stellar core with a mass above the neutron star limit. There, the internal energy associated with the core collapse becomes so great that it exerts a significant gravitational force of its own. This extra gravity sets in motion a positive feedback loop, in which the continuing collapse releases even more energy, which makes gravity even stronger, which releases more energy that increases the strength of gravity, and so on. To the best of our knowledge, nothing can stop this positive feedback. The core must collapse without end, becoming a black hole.

BLACK HOLE MASSES

The black hole formation process that we've discussed might seem to imply that all black holes would have masses of a few to a few tens of times the mass of the Sun, since those are the masses we might expect for collapsing stellar cores above the neutron star limit. However, as we discussed in chapter 1, a second major category of black holes is known to exist: supermassive black holes found at the centers of galaxies.

Scientists are not sure exactly how these supermassive black holes formed, but it's easy to envision ways in which it might have happened.

For example, because they are located in the dense centers of galaxies, they might have started simply through the merger of many black holes formed in supernovae. Once their masses grew large enough, their tidal forces would have become strong enough to rip apart other stars that passed near them. The gas from those stars would then form an orbiting disk around the black hole, and the friction within this disk would gradually cause the orbits of individual particles to decay until they fell into the black hole, adding to its mass. In any event, regardless of the precise mechanism, there's really nothing surprising about the existence of supermassive black holes. After all, if gravity is strong enough to overcome all sources of pressure even in objects with as little as about 3 solar masses of material, it can certainly overcome the pressure in much more massive objects.

You should now understand why, back in chapter 1, I stated that as far as we know, all black holes have masses at least a few times the mass of our Sun. The gravity of lower-mass objects simply isn't strong enough to overcome all the forms of pressure that can fight back against it. Nevertheless, some physicists have suggested that there might be cases in which processes besides gravitational collapse could create "mini black holes" with much smaller masses.

Two general types of mini black holes have been proposed. The first are mini black holes that might have formed during the Big Bang. The basic idea is that the tremendous energy of the Big Bang might have generated forces that could have helped squeeze objects enough to make them black holes, even if their own gravity wouldn't have been strong enough to do it by itself. If true, the universe could be populated by numerous black holes with masses similar to those of planets or small stars. This possibility has been investigated by scientists who try to model conditions during the early moments of the Big Bang. Although we cannot fully rule it out, most of the models suggest that relatively little mass, if any, would have been turned into mini black holes. Observational searches for such black holes (by looking for the gravitational lensing they would cause on stars behind them) seem to confirm their nonexistence.

The second type of mini black hole is hypothesized to be far smaller, produced by some type of quantum fluctuation occurring at a subatomic scale. These potential "micro black holes" gained notoriety when the media reported fears that they might be created in the Large Hadron Collider in Europe, and that they might then go on to destroy our planet. Some physicists have indeed proposed scenarios in which such micro black holes could be produced in the Large Hadron Collider, but even if they are right, there's nothing to worry about. The reason is that while the Large Hadron Collider can generate particles from greater concentrations of energy than any other machine that humans have ever built, nature routinely makes such particles. Some of those particles must occasionally rain down on Earth, so if they were dangerous, we would have suffered the consequences long ago.

In case you are wondering how a micro black hole could be "safe," the most likely answer has to do with a process called *Hawking radiation*, which gets its name because it was first proposed by renowned physicist Stephen Hawking. The details are fairly complex, but in essence Hawking showed that the laws of quantum physics imply that black holes can gradually "evaporate" in the sense of having their masses decrease, even while nothing ever escapes from within their event horizons. The rate of evaporation depends on a black hole's mass, with lower-mass black holes evaporating much more rapidly. The result is that while the evaporation rate would be negligible for black holes with star-like masses or greater, micro black holes would evaporate in a fraction of a second, long before they could do any damage.

OBSERVATIONAL EVIDENCE FOR BLACK HOLES

Now that we understand black hole formation, the observational search for black holes becomes fairly straightforward. We're simply looking for objects that are extremely dense, but too massive to be neutron stars.

FIGURE 7.3 This painting shows an artist's conception of the Cygnus X-1 system. Gas from the companion star (right) swirls toward the black hole, and its high temperature means it emits intense X-rays. We can be confident that the compact central object is a black hole, because its mass is too large to be a neutron star. The inset shows the system's location in the constellation Cygnus. Painting by Joe Bergeron.

As we discussed in chapter 1, the easiest way to find such objects is to look for sources of intense X-rays. Recall the case of Cygnus X-1, in which we observe X-rays coming from very hot gas swirling around a compact object in a binary star system (figure 7.3). From the speed at which the hot gas orbits the compact object, we conclude that the object must be either a neutron star or a black hole. Because the object's mass exceeds the neutron star limit by a fairly wide margin (15 solar masses, versus the 3-solar-mass limit), we conclude that Cygnus X-1 contains a black hole.

The case is even stronger for supermassive black holes at the centers of galaxies. Recall that some of these black holes have millions to billions of times the Sun's mass packed into very small volumes of space. These masses are so far above the neutron star limit that it's difficult to conceive of any way that these objects could be anything other than black holes.

SINGULARITY AND THE LIMITS OF KNOWLEDGE

As we've seen, the evidence strongly supports the idea that black holes are both real and common in the universe. And that brings us back to the question of what lies inside them.

From a scientific standpoint, this question is very difficult to answer. The problem is that the event horizon marks another important boundary besides being the boundary between the inside and outside of the black hole. Because we cannot observe anything within the event horizon, there is no way to gather either observational or experimental evidence about what exists within it. In that sense, the inside of a black hole lies outside the realm of science, just as it lies outside of our observable universe.

Nevertheless, we can use the laws of physics to predict what *should* happen inside a black hole. While we should keep in mind that there is no way to check whether these predictions are correct, they can still provide interesting results that might point to other ideas that *can* be tested.[7] With that caveat, let's return to an imploding stellar core that is forming a black hole.

Because nothing can stop the crush of gravity in this imploding core, it seems logical to conclude that it will continue to collapse until it becomes an infinitely tiny and dense point known as a *singularity*. In other words, all the original mass of the object that became a black hole should be compressed to infinite density at the singularity, a place where spacetime becomes infinitely curved. Similarly, at least from his point of view, the fate of your colleague who plunged into the black hole on your trip would have been to fall rapidly to the singularity where he, too, would be crushed to infinite density.

---

7. This idea is not really so different from the way we study other objects. For example, we cannot directly sample Earth's core or the interior of the Sun, but we are confident that we understand them because we can calculate the properties they must have in order to explain the observations we make from outside them.

Unfortunately, while the idea of an infinitely dense singularity seems to make sense according to general relativity, it does not make quite as much sense according to another very successful theory of physics: the theory of quantum mechanics, which explains the nature of atoms and of subatomic particles. Without going into too much detail, we can state the basic problem as this: Quantum mechanics includes the famous *uncertainty principle*, which tells us that we cannot know both an object's location and its motion with perfect certainty. As a result, quantum mechanics in essence tells us that a singularity should be a point at which spacetime would fluctuate chaotically, which is not the same thing as being a point of infinite spacetime curvature.

The fact that general relativity and quantum mechanics give different answers about the nature of the singularity means that these two answers cannot both be correct. We have therefore run up against the limits of current scientific knowledge. While that may be unfortunate if you'd like to understand singularity right now, scientifically it is very exciting. In essence, the situation is analogous to what scientists faced back when the equations of electromagnetism did not provide a reference frame for the speed of light, or when the orbit of Mercury did not quite match what Newton's laws predicted. Just as those problems helped lead Einstein to discover his special and general theories of relativity, scientists are optimistic that the problem of singularity may ultimately point us to a new and better theory of nature, one that will supplant both general relativity and quantum mechanics in much the same way these theories supplanted our earlier theories of gravity and of atoms.

HYPERSPACE, WORMHOLES, AND WARP DRIVE

The fact that we cannot observe the inside of black holes makes them ripe territory for science fiction. For example, in chapter 1 we briefly discussed the idea that black holes might provide passage from one

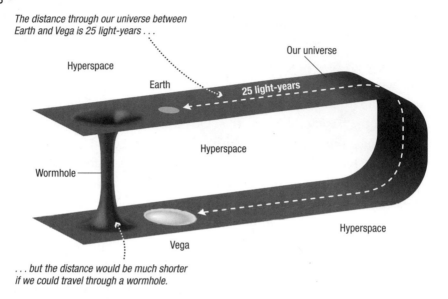

The distance through our universe between Earth and Vega is 25 light-years . . .

. . . but the distance would be much shorter if we could travel through a wormhole.

FIGURE 7.4 The surface of the rubber sheet in this diagram represents the space between Earth and the star Vega. The distance through space from Earth to Vega is 25 light-years, but a wormhole might conceivably offer a much shorter path. Adapted from an illustration by Kip Thorne in his book *Black Holes and Time Warps: Einstein's Outrageous Legacy* (Norton, 1994).

part of the universe to another. It should now be possible to see where this idea comes from. If you visualize a piece of the universe as a curved rubber sheet, with black holes as deep "wells" in the sheet, you might imagine that two black holes could connect to each other somewhere out in hyperspace, meaning in places that lie outside of our ordinary four dimensions of spacetime. This is the basic idea behind *wormholes* (figure 7.4), though mathematical calculations with Einstein's equations indicate that they don't actually connect two black holes. Unfortunately, while the mathematics suggests that wormholes should really exist, it also indicates that they would be unstable and would collapse the instant you tried to travel through them.

Nevertheless, some physicists (especially Kip Thorne of the California Institute of Technology [Caltech]) have investigated possible ways of getting around the instability problem, imagining ways in which an advanced civilization might use wormholes to create an interstellar tunnel system. So far, none of these solutions seems particularly likely to be viable, but we can't rule wormhole travel out completely. That is why wormholes are so popular in science fiction, and why Carl Sagan used a wormhole tunnel network in his terrific novel *Contact*, which was later made into a movie of the same name.

Carrying the idea of tunnels through hyperspace further, you can probably see other ways in which general relativity might seem to allow for faster-than-light travel from one place in the universe to another. For example, the writers of the *Star Wars* movies imagined that you might simply "jump" into hyperspace, effectively leaving our universe, then jump back in at a place of your choosing. In *Star Trek*, the writers created warp drive, which presumably is a mechanism that warps or folds spacetime to bring distant points into hyperspace contact, allowing you to move rapidly between them. Note that none of these ideas violates special relativity's prohibition against faster-than-light travel, because that prohibition applies only to travel *through* ordinary space. If you leave the universe to travel through hyperspace, that prohibition need no longer apply.

These are all fun ideas to think about, and for the time being, the known laws of physics do not rule out any of these exotic forms of travel. But even if any of them are possible in principle, actually using them for practical travel remains well beyond any technology we can envision. After all, it's not exactly clear how you'd go about jumping in and out of our universe to pass through hyperspace, and finding a way to warp spacetime to a greater extent than it is naturally warped by gravity seems a rather steep engineering challenge. In addition, many scientists have voiced another objection to these ideas: Because of the way that space and time are intertwined in spacetime, all of these possible modes of travel would seem to allow for travel through time as well as space. The well-known paradoxes of time travel, such as traveling into the past and preventing your parents from meeting, make many physicists doubt that time travel is possible. In the words of Stephen Hawking, time travel should be prohibited "to keep the world safe for historians."

Still, the bottom line is that neither time travel nor travel through hyperspace can yet be ruled out with the same confidence that we can rule out the possibility of exceeding the speed of light in ordinary space. Until we learn otherwise, the world remains safe for science fiction writers who choose their fictional space travel techniques with care, avoiding any conflicts with relativity and other known laws of nature.

BLACK HOLES DON'T SUCK

We have now come full circle on the imaginary journey to a black hole that we began in chapter 1. So as we wrap up our discussion of black holes, let's step back and summarize the key lessons that we have learned about what they are, how they affect objects around them, and what happens to objects falling into them.

What they are: Einstein's general theory of relativity tells us that what we view as gravity actually arises from curvature of spacetime, and that this curvature is created by masses. Black holes are masses that have shrunk to such small size that they have essentially created holes in the observable universe. Once an object falls into a black hole, the outside universe loses all contact with it.

What happens to objects around them: The gravity of a black hole is no different than the gravity of any other object of the same mass, except that it becomes much more extreme—meaning much greater curvature of spacetime—when you approach the black hole very closely. At a distance, you'll orbit a black hole just as you would any other larger mass. It will not suck you in.

What happens to objects falling into them: First, it's difficult to fall into a black hole by accident, because black holes are so compact in size that you'd need nearly perfect aim to hit one from afar. About the only thing that falls into a black hole easily and naturally is gas that is in its vicinity, and this happens because swirling gas creates friction that can cause the gas particle orbits to decay until they fall into the black hole. If you watch something falling into a black hole, from the outside you'd perceive its time coming to a stop as it approaches the event horizon. At the same time, you'd also see its light becoming infinitely redshifted, which means it will disappear from your view. The redshift explains why material falling into a black hole will indeed disappear from our view relatively quickly, and yet we'll never actually see it cross the event horizon.

With that summary behind us, I'd like to end this chapter with a personal observation about what black holes tell us about the nature of science. I've often heard nonscientists argue that science is somehow limiting, and that scientists are so skeptical that they are closed-minded to new ideas. The story of black holes provides a powerful counterargument. Before Einstein, anyone proclaiming that there could be holes in the universe bounded by event horizons where time comes to a stop and light becomes infinitely redshifted would probably have been considered to be crazy. Even after Schwarzschild used Einstein's equations to show that they allowed for the existence of the things we now call black holes, nearly all scientists assumed that they were still too strange to be true. As recently as the 1960s, any poll of scientists would likely have found most of them assuming that some undiscovered law of nature would ensure that such strange objects could not really exist. Today, that situation has been completely reversed, and it's difficult to find any physicist or astronomer who doubts that black holes are both real and common in the universe.

This dramatic change in scientific outlook is a direct result of the evidence-based nature of science. No matter how strange any idea may seem at first, if the evidence becomes strong enough, scientists will ultimately come to accept it. That is why my personal favorite definition of science is that it is *a way of using evidence to help us come to agreement.*

Whatever controversial new idea may come along—whether it is the idea that Earth orbits the Sun, that life evolves through time, or that gravity arises from curvature of spacetime—science provides the *only* means by which we can ever come to agreement about whether the idea is correct or destined for the dustbin of history.

8

THE EXPANDING UNIVERSE

EINSTEIN'S THEORIES sound so revolutionary and modern that it's easy to forget that he discovered them at a time when much of the rest of our understanding of the universe was still quite limited. For example, as we discussed in chapter 7, while general relativity made allowance for black holes, few people thought they might really exist until many decades later. Similarly, although $E = mc^2$ suggested that stars could in principle be shining by converting a small fraction of their mass into energy, the mechanism of nuclear fusion wasn't discovered until more than 30 years after Einstein first came up with the equation in his special theory of relativity.

Perhaps most surprisingly to people today, the human conception of the universe was far different when Einstein did his work in relativity. Today, elementary school kids can tell you that the Milky Way is the galaxy we live in, and that it is just one of a great many galaxies in the universe. But when general relativity was published in 1915, astronomers were still actively debating whether separate galaxies existed,[1] with many (perhaps most) coming down on the side of the Milky Way

---

1. This fact often surprises people, since today we can easily photograph other galaxies with telescopes. But the telescopes of the time were not yet powerful enough to see galaxies as much more than somewhat fuzzy patches of light, and it therefore was not obvious whether these patches were clouds of gas within the Milky Way or separate collections of stars.

representing the entire universe. Given that we now know the observable universe to contain some 100 billion galaxies, this means that the universe of 1915 was thought to be about 100 billion times smaller (in number of galaxies) than we know it to be today.

With this historical backdrop, we now move to our final topic for this book: a prediction made by general relativity that seemed so incredible at the time that even Einstein himself did not believe it. As we will see, despite Einstein's mistake in disbelieving his own theory, he came up with an idea that may yet prove important to our understanding of the universe.

EINSTEIN'S BIGGEST BLUNDER

Shortly after Einstein published the general theory of relativity, he was working with its equations when he realized that they had a rather troubling implication: Because all matter attracts all other matter through gravity, his equations suggested that the universe could not be stable. That is, when he tried to assume a universe in which all objects stayed in their places, he found that gravity would pull them all together, causing the universe to collapse. In essence, his theory seemed to predict that the universe should have long ago collapsed into its own black hole.

With hindsight, we can see that there are at least two possible ways to reconcile general relativity with a universe that has not collapsed out of existence. The first is to assume that general relativity is correct as it stands, and that the reason the universe has not collapsed is because it is expanding. In other words, if we assume that we live in an expanding universe, then the expansion would offset the tendency for gravity to make the universe collapse. The second way to reconcile the theory with reality is to assume that the theory is missing something— in particular, that the equations are missing some term that would offset the overall attractive force of gravity. In that case, we might try to "fix" general relativity by adding some new term to the equations that would make the universe hold still.

Einstein, of course, did not have the benefit of our hindsight. Moreover, for reasons that are not completely clear, Einstein believed that the universe should be static and eternal. Therefore, only the second path to reconciling his theory with reality seemed open to him, and that is the path he chose. In essence, he added a fudge factor to his equations of general relativity for the sole purpose of counteracting the normal attractive force of gravity, so that the universe would fit his personal conception of it. This fudge factor appears as a single term in the equations, and Einstein called it the *cosmological constant*.

Einstein introduced the cosmological constant to the world in a paper he published in 1917. Even at that time, he seemed almost apologetic about adding the term, acknowledging that it had no evidence-based justification and that it complicated the otherwise simple structure of his equations. He also recognized that his equations might be fine as they were if not for his insistence on making the universe static and eternal.

Einstein later called his introduction of the cosmological constant the "biggest blunder" of his career. Because the comment comes to us secondhand (from an autobiography by physicist George Gamow), we don't know exactly why Einstein made it, but we can make a reasonable guess. Einstein had many deeply held beliefs, but he also prided himself on his devotion to the evidence-based processes of science. For example, in the case of general relativity, he called the key idea (the equivalence principle) his "happiest thought," but it was his theory's later success in explaining the precession of Mercury's orbit that he considered the high point of his scientific career. The introduction of the cosmological constant appears to be a rare case in Einstein's career in which he adjusted a theory to fit a preconceived idea about the universe without looking for confirming evidence. In that context, he may have perceived the cosmological constant as his "biggest blunder" not only because it prevented him from predicting the expansion of the universe, but perhaps also because he had allowed himself to stray from his devotion to the principles of science.

EXPANSION DISCOVERED

Putting aside Einstein's cosmological constant, we're left with a remarkable fact: Because the universe clearly is not collapsing, the general theory of relativity actually *predicts* that the universe should be expanding. This prediction was verified by Edwin Hubble a little more than a decade after Einstein's publication of general relativity.[2] It has since been verified by so many observations that we now regard the expansion of the universe as an established fact.

Hubble's discovery of the expanding universe was the product of a decade of careful observations by himself and others, but it really consisted of two key parts. First, he established the existence of galaxies beyond the Milky Way. He did this by having access to a new, powerful telescope that enabled him to see individual stars and star clusters in relatively nearby galaxies. By doing so, he could estimate the distances to the galaxies, and the distances were so vast that everyone agreed the galaxies had to lie beyond the Milky Way. Next, he began making distance estimates for many galaxies, while also measuring the speeds (by looking for shifts in spectral lines) at which the galaxies were moving toward or away from Earth. He discovered that with the exception of some very nearby galaxies, all other galaxies are moving away from us, and the farther away they are, the faster they are moving.

You can understand how that observation leads to the conclusion that we live in an expanding universe by thinking about a raisin cake baking in an oven (figure 8.1). Imagine that you make a raisin cake in which you carefully place the raisins so that the distance between adjacent raisins is always 1 centimeter. You place the cake in the oven, and

---

2. A Belgian priest and scientist named Georges Lemaître actually published a paper about the expanding universe two years before Hubble's own published work. Some historians argue that Lemaître should therefore be given credit for the discovery. However, Hubble may have been unaware of Lemaître's paper, which was published in French, and recent sleuthing work by Mario Livio of the Hubble Space Telescope Science Institute indicates that Lemaître himself did not presume to deserve credit for the discovery.

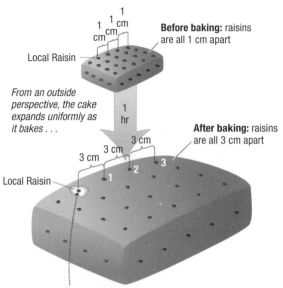

FIGURE 8.1 If you lived in a raisin inside an expanding raisin cake, you would observe all other raisins to be moving away from you, with more distant raisins moving faster. In the same way, the fact that we observe more distant galaxies to be moving away from us at higher speeds implies that we live in an expanding universe.

over the next hour it expands until the distance between adjacent raisins has increased to 3 centimeters. The fact that the cake has expanded in size will be pretty obvious as you look it from the outside. But what would you see if you lived *in* the cake, as we live in the universe?

To answer this question, we can pick any raisin (it doesn't matter which one) and identify it as our "Local Raisin." Figure 8.1 shows one possible choice for the Local Raisin, with several other nearby raisins labeled, both before and after baking. The accompanying table summarizes what you would see if you lived within the Local Raisin. Notice, for example, that Raisin 1 starts out at a distance of 1 centimeter before baking and ends up at a distance of 3 centimeters after baking, which means it moves a distance of 2 centimeters away from the Local Raisin during the hour of baking. Hence, its speed as seen from the Local Raisin is 2 centimeters per hour. Raisin 2 moves from a distance of 2 centimeters before baking to a distance of 6 centimeters after baking, which means it moves a distance of 4 centimeters away from the Local Raisin during the hour. Hence, its speed is 4 centimeters per hour, or twice as fast as the speed of Raisin 1. As the table shows, the pattern continues, so that you'd see all other raisins moving away from your location in the Local Raisin, with more-distant raisins moving away from you faster. This is exactly what Hubble observed for galaxies, and it is what allows us to conclude that we live in an expanding universe.

The main problem with the raisin cake analogy is that the cake is a three-dimensional object sitting within a larger three-dimensional space. This means that we see a center and edges when we look at the cake from the outside, and that it expands into pre-existing space. According to general relativity, the structure of the universe is *defined* by the masses within it, which means that we cannot think of space or spacetime independently of the universe. In more practical terms, this means that the universe does not have either a center or edges, and that it does not expand into a pre-existing space. Rather, the existing space between galaxies essentially gets stretched out as the universe expands.

This fact inevitably leads people to ask how the universe can be expanding without expanding *into* something. As usual, our only hope of visualizing it is with a two-dimensional analogy. In this case, let's imagine the universe as the *surface* of an expanding balloon; like Earth's surface, the balloon's surface is two-dimensional because

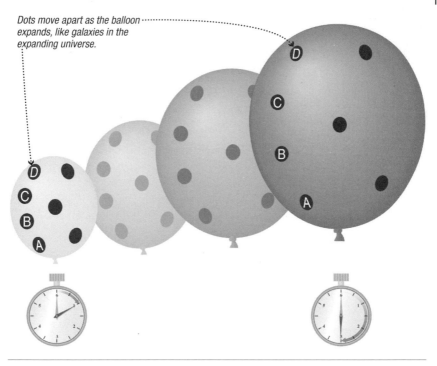

Dots move apart as the balloon
expands, like galaxies in the
expanding universe.

FIGURE 8.2 The *surface* of an expanding balloon provides a good two-dimensional analogy to the expansion of our universe. Note that only the surface represents the universe; regions inside and outside the balloon are not part of the universe in this analogy.

there are only two independent directions of motion possible on it (such as north-south and east-west). We are therefore using the two-dimensional surface to represent all three dimensions of space, which means that in this analogy, regions inside the balloon and outside its surface *are not part of our universe.*

Figure 8.2 shows dots, representing galaxies, on an expanding balloon. Just as in the raisin cake, you can pick any dot to be your Local Dot, and you'll see all other dots moving away from you as the balloon expands, with more-distant dots moving away from you faster. This time, however, the analogy matches several key features of the real universe. Like the surface of Earth, the surface of the balloon has no center and no edges. (Of course, there is a center inside the balloon, but the inside is not part of the surface; in other words, just as New York is no more "central" to Earth's surface than any other city, no dot on the balloon's surface is any more centrally located than any other dot.) Moreover, as the balloon expands, the surface grows in size, but it does not grow into some pre-existing piece of balloon. Rather, it is simply the surface itself that is stretching out as the balloon expands, in the same way that space stretches out as our universe expands.

THE BIG BANG

The balloon analogy also leads us to another prediction. If we extrapolate backward, the balloon's surface must have been smaller and smaller as we go further back in time. At some point, it would have been infinitesimally small, and you can't get any smaller than that. We infer that there must have been a starting point for the balloon's expansion, and by analogy we predict that there must similarly have been a starting point for the universe's expansion. In other words, the fact that our universe is expanding leads us to predict that the universe must have been born at some particular moment in the past, which is the moment at which the expansion began. We call that the moment the *Big Bang*. Given how rapidly the universe is expanding, we can work the expansion backward to estimate when the Big Bang occurred. The best current estimate puts the age of the universe at a little less than 14 billion years. (More precisely, data released in 2013 from the European Space Agency's *Planck* mission put the age of the universe at about 13.8 billion years.)

Before we continue discussing expansion, I want to focus briefly on two crucial points about the Big Bang. First, as our analogy shows, the Big Bang is simply a name given to the beginning of the expansion; it is *not* an explosion that sent matter flying into some pre-existing space, because there was no pre-existing space. Second, the idea of the Big Bang is a logical prediction made from the observed fact of an expanding universe, which is in turn confirmation of general relativity's prediction that the universe cannot be standing still. Nevertheless, like anything in science, a prediction remains speculation until evidence is found to support it. In the case of the Big Bang, scientists have indeed found strong evidence that it really occurred.

In brief, there are three major lines of evidence that support the idea of the Big Bang. First, remember that because light takes time to travel vast distances through space, we see objects at great distances as they were a long time ago. For example, when we look at galaxies that are 7 billion light-years away,[3] we are seeing light that has traveled through space for 7 billion years to reach us, which means we are seeing the galaxies as they looked 7 billion years ago. If there really was a Big Bang some 14 billion years ago, then such galaxies should (on average) be only about half the age of galaxies that are near us. Distant galaxies do indeed show clear evidence of being younger than nearby galaxies, which supports the idea of the Big Bang because it implies a finite age for the universe.

The second key line of evidence for the Big Bang comes from observations of the *cosmic microwave background*, which refers to microwave radiation that specialized telescopes have detected coming from all directions in space. To understand how this supports the Big Bang, think about what happens when you compress air: The compression makes the air become hotter. In much the same way, the theory of the Big Bang predicts that the universe should have been hotter when it was younger, because any piece of the universe was in essence compressed to a smaller size. Hot objects always emit radiation, so the young universe should have been filled everywhere with intense light. As the universe has expanded and cooled, the stretching of space should have

---

3. Large distances are somewhat ambiguous in an expanding universe, since distant galaxies today must be farther away from us than they were at the time that the light we see from them started traveling toward us. In this book, when I state a distance such as 7 billion light-years, what I really mean is a galaxy located at a distance such that its light has taken 7 billion years to reach us. To avoid this type of ambiguity, astronomers often say that the galaxy is located at a "lookback time" of 7 billion years, because we see it as it was 7 billion years ago.

gradually stretched out the wavelengths of this light. Calculations first made in the 1940s suggested that *if* there had been a Big Bang, the universe today should be filled with radiation characteristic of an overall temperature a few degrees above absolute zero, which means it would be detectable as microwave radiation. The cosmic microwave background, first detected in the early 1960s, has a temperature of about 3 degrees above absolute zero, consistent with the prediction made by the Big Bang theory. In fact, more detailed analysis of the Big Bang theory predicts precise characteristics for the spectrum of the cosmic microwave background, and the observations match these predictions with fantastic precision.

The third line of evidence comes from the observed overall chemical composition of the universe. The Big Bang theory can be used to calculate the temperature and density of the early universe, and these conditions can in turn be used to predict the chemical composition of the early universe. Very early on, the only element would have been hydrogen—whose nuclei are simply individual protons—because it was too hot for individual protons and neutrons to be held together in larger atomic nuclei. However, for a remarkably short period of about 5 minutes after the Big Bang, it should have been possible for some nuclear fusion to occur, and the calculations predict that the chemical composition of the universe should have been transformed to become 75% hydrogen and 25% helium (by mass). Moreover, with the exception of the relatively small fraction of this material that has since been fused into heavier elements by stars, we would expect the universe to still have this same basic chemical composition today—and observations show that it does. In other words, the Big Bang theory predicts the observed chemical composition of the universe.

To summarize, the idea of a Big Bang is a natural outgrowth of general relativity's "prediction" of an expanding universe. With three strong lines of evidence supporting the idea of the Big Bang, there seems little scientific doubt that the expansion of our universe really did have a beginning, approximately 14 billion years ago.

THE GEOMETRY OF THE UNIVERSE

Our discussion of the balloon analogy may make you wonder about the overall "shape" of spacetime. We know that gravity arises from curvature of spacetime, and that locally the shape of spacetime can take many different forms. However, spacetime as a whole must have some overall curvature that is the result of the combined action of all the masses within it. That is, every individual mass within the universe causes some local curvature, and together all those local curvatures will add up to some global shape. The idea is similar to the way we view Earth: Locally, Earth's surface is curved in many different ways by mountains, valleys, and other geographic features, but globally our planet is clearly round.

When we use the balloon analogy, we are essentially assuming that all the local curvatures add up to something that ends up curving back on itself like the surface of Earth. That is not the only possibility, however. A second possibility is that the overall shape of spacetime looks rather like a flat trampoline, just pockmarked with local depressions of gravity. A third possibility is that instead of curving back on itself like Earth or a balloon, spacetime spreads outward rather like the surface of a saddle.

Figure 8.3 uses two-dimensional surfaces to show all three of the possible geometries. Note that, to avoid having a center and edges, you have to imagine the flat and saddle-shaped geometries extending to infinity; only the balloon-like, spherical geometry has a finite surface. To complete the visualization, you need to imagine all three surfaces expanding to represent an expanding universe, and as always remember that these surfaces are only two-dimensional analogies to the structure of space in four-dimensional spacetime.

General relativity does not tell us which of these three possible geometries the universe actually has. To learn that, we must approach the question in other ways. There are two major approaches. One is to try to determine the total density of matter and energy in the universe. This should tell us about the geometry, because greater density means

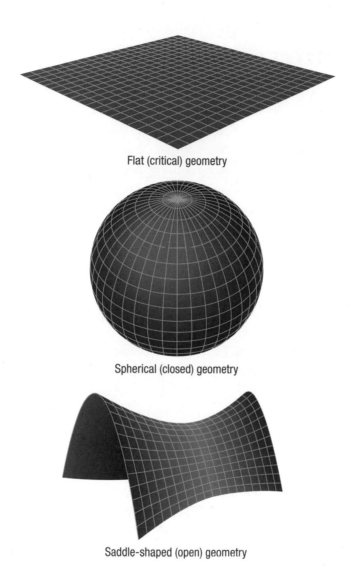

Flat (critical) geometry

Spherical (closed) geometry

Saddle-shaped (open) geometry

FIGURE 8.3 Two-dimensional analogies to the three possible overall geometries for the universe.

stronger gravity and hence greater curvature, so we can use the density to calculate the overall geometry. Alternatively, we can examine the way in which the rate of expansion has changed with time, because that would also tell us the overall strength of gravity for the universe as a whole.

PATTERNS OF EXPANSION

If we think logically about it, we would expect that gravity would gradually slow the expansion of the universe with time. If the universe contains enough mass—and hence enough gravity—then the expansion would eventually come to a halt and then reverse. In that case, the universe might someday end in a "Big Crunch." If the universe contains somewhat less total mass, then we'd still expect gravity to slow the expansion gradually, but never to slow it enough for the expansion to halt and reverse. The universe would in a sense just fade away over time, as stars eventually burn out and galaxies go dark.

These ideas suggest a simple way to determine which possibility is correct: Measure how the expansion rate is changing with time. If the expansion is someday going to stop, then it should already be slowing at a fairly significant rate. If the expansion is going to continue forever, then the rate of slowing should be much smaller.

How do we go about measuring how the expansion rate has changed with time? In principle, it's easy. Again, because light takes time to travel the vast distances across the universe, we see more distant objects as they were when the universe was younger. For example, when we look at galaxies within, say, 200 million light-years, we can use the speeds at which they are moving away from us to determine what the expansion rate has been during the past 200 million years. If we then look to galaxies that are, say, 7 billion light-years away, we can use the speeds at which these galaxies are moving away from us to determine what the expansion rate was 7 billion years ago, when the universe was only half its current age.

In practice, measuring the expansion rate for different times in the past is very difficult (primarily because it is difficult to measure galaxy distances with sufficient precision). Nevertheless, starting in the 1990s, the Hubble Space Telescope and other powerful new observatories gave astronomers the ability to start making such measurements. The results, first announced in 1998, came as an almost complete shock to astronomers.

The shock was this: As we just discussed, almost everyone naturally assumed that gravity would be slowing the expansion, and the only question was whether it was slowing it a lot or a little. However, the observations show that the expansion is not slowing at all. Instead, the expansion is *accelerating* with time.

NOT A BLUNDER AFTER ALL?

What could possibly be causing the expansion of the universe to accelerate rather than to slow with time? The truth is, no one knows. This fact sometimes comes as a surprise to people who follow astronomy in the news, because most scientists have already agreed on a name for the answer: *dark energy.* You'll therefore hear scientists talking about the "fact" that dark energy is causing the acceleration of the universe, and perhaps even discussing some presumed characteristics of dark energy, which means characteristics inferred from the way in which the expansion is accelerating. But having a name for something does not necessarily mean that we understand it, and if there's one thing I want you to understand about dark energy, it is this: Every idea you've ever heard about dark energy is little more than a guess, or at best an educated guess. It doesn't matter who the idea comes from, or how prominent a scientist he or she may be. For the time being, we lack any conclusive evidence about the nature of dark energy, which means we simply do not know what it is.

The quest to understand dark energy is one of the greatest adventures in science today, but for our purposes in a book about relativity we'll focus only on a remarkable side story. When we ask whether the equations of general relativity are consistent with the idea of an accelerating expansion, the answer we find is this: Mathematically, general relativity is compatible with an accelerating expansion if we include a term that can counteract ordinary gravity. In other words, it works if we include Einstein's cosmological constant, the fudge factor that he called his "biggest blunder."

Whether this fact has any bearing on reality remains to be seen. After all, it's also possible that, just as we know that relativity and quantum mechanics give different answers at the singularity of a black hole, relativity may not necessarily give the correct answer for the overall geometry of the universe. That is, while general relativity has passed many precise tests of its validity over localized regions of the universe, and in some cases over very wide swaths of the universe, we cannot yet be sure that it is the full story for the universe as a whole. Still, it's fun to ponder the idea that Einstein, even in what he considered to be his worst scientific moment, may yet prove to have been ahead of his time.

THE FATE OF THE UNIVERSE

Regardless of its cause, the observed acceleration of the expansion has implications for the ultimate fate of the universe. Remember that before the acceleration was discovered, there seemed to be two possible fates for the universe: an end in a Big Crunch, or a gradual but never-ending slowing of the expansion. Assuming that future observations continue to support the case for an accelerating expansion, we must now introduce a third possibility: that the expansion will continue to accelerate. Some scientists have even suggested that an accelerating expansion might eventually push the expansion rate so high that the universe would someday tear itself apart in what has been called the "Big Rip," though this idea is controversial at best.

A more direct implication of an accelerating expansion is that it would seem to rule out the possibility of the expansion ever stopping and reversing, implying that the expansion of the universe will continue forever. Indeed, the idea that an accelerating expansion implies a permanent expansion seems so logically obvious that many people take it as a given. However, as I've emphasized throughout this book, logic is not enough in science. Until we actually understand what is causing the acceleration—which means being able to conduct observational or experimental tests that verify the presumed cause—we cannot be confident that our logic is correct.

More important, even if we discover the source of the mysterious dark energy that drives the acceleration, and even if this discovery supports the logic of a permanent expansion, there's still an important caveat. Namely, our logical extension of today's acceleration to tomorrow's fate of the universe can be no more valid than the best of today's scientific knowledge. In the particular case of the fate of the universe, remember that the mere idea of acceleration came as a huge surprise to scientists when it was discovered less than two decades ago. All it would take to change our view of the fate of the universe again is another equally surprising discovery, and such a discovery could in principle be made at any time between now and the end of time.

EINSTEIN'S LEGACY

We began this book with an imaginary voyage to a black hole. Our quest to understand what we experienced on that voyage has led us through an introduction to Einstein's special and general theories of relativity, which in turn has led us to pondering the history of the universe from its beginning to its possible ends. Unless you already knew something about this subject, I suspect that you've been very surprised to learn of these connections between the relativity of motion and the overall nature of space, time, and the universe.

Einstein's legacy is usually discussed in terms of his discoveries, and there's no doubt that he revolutionized physics and our understanding of the universe. He taught us that space and time are inextricably linked, he gave us a new way to understand gravity, and his theories are now used to understand topics ranging from exotic objects like black holes to the overall geometry of the universe.

To me, however, his greatest legacy comes in the way he demonstrated the incredible power of scientific thought. As a teenager, Einstein began to wonder what the world would look like if he could ride on a beam of light. But he didn't stop there; instead, he took the initiative to learn mathematics and physics at a deep enough level to investigate the question quantitatively, and to explore where different avenues of thought would lead. This is the essence of science, and I hope that Einstein's achievements will inspire many more people to recognize the value of science and to put its power to work in helping us to both understand the world and make it a better place for all of us.

YOUR INDELIBLE MARK
ON THE UNIVERSE

I BEGAN this book with the claim (in the preface) that relativity is *important* to understanding how we as humans fit into the overall scheme of the universe. Now that we have completed our introduction to Einstein's theories, it seems a good time to look back and think more deeply about that claim. Of course, different people may come to different conclusions about exactly what makes relativity important, and I encourage you to come up with your own ideas. For myself, however, I find relativity to be important on at least four different levels.

The first level is that of pure science. In the more than 100 years since Einstein first introduced relativity to the world, both his special theory and his general theory have been extensively and repeatedly tested. Today, there can be no doubt as to their validity, at least within the realms in which they have been tested, and we therefore cannot understand nature without first understanding relativity. To review just a few examples: We cannot understand how stars shine without first understanding $E = mc^2$; we cannot understand what a black hole is until we first understand gravity as arising from curvature of spacetime; we cannot understand how the universe can be expanding without expanding "into" something unless we first understand the possible four-dimensional geometries of spacetime as a whole; and our GPS units would not work without the calculations of relativity. Indeed, relativity is now as integral to our overall understanding of the universe as the idea that Earth is a planet orbiting the Sun or that gravity makes objects fall to the ground.

The second level on which I find relativity important is that of our perception of reality. Our common experiences make us grow up with the presumption that space and time are separate and distinct, but relativity has shown us otherwise. As Einstein's colleague Hermann Minkowski said in 1908, "Henceforth space by itself, and time by itself, are doomed to fade away into mere shadows, and only a kind of union of the two will preserve an independent reality." Moreover, general relativity alters our perception of gravity, changing it from Newton's absurd action at a distance to a natural consequence of the geometry of a spacetime that is curved by masses within it. These changes in perception may not have much effect on the way we go about our daily lives, but they certainly should change the way we understand and interpret the universe around us.

The third level of importance lies in what I think Einstein's discovery of relativity tells us about our potential as a species. The science of relativity may seem disconnected from most other human endeavors, but I believe that Einstein himself proved otherwise. Throughout his life, Einstein argued eloquently for human rights, human dignity, and a world of peace and shared prosperity. His deeply held belief in underlying human goodness is all the more striking when you consider that he lived through both World Wars, that he was driven out of Germany by the rise of the Nazis, that he witnessed the Holocaust that wiped out more than six million of his fellow Jews, and that he saw his own discoveries put to use in atomic bombs. No one can say for sure how he maintained his optimism in the face of such tragedies, but I see a lesson in relativity. As you've seen, the ideas of relativity seem so strikingly counter to the "common sense" that we grow up with that at first they are difficult to believe. Indeed, I suspect that for much of human history, relativity would have been rejected out of hand, just because it seemed so outrageous. Yet we live in a time when, thanks to the process that we call *science*, evidence is now considered to be more important than preconceptions. We have come to accept relativity because the evidence supports it so strongly, even though it has forced us to redefine our perception of reality. To me, this willingness to make judgments based on evidence shows that we are growing up as a species. We have not yet reached the point where we always show the same willingness in all our other endeavors—if we had, there would be no more injustice or corruption in the world—but the fact that we've done it for science suggests we have the potential.

Finally, I find relativity to be important on a fourth and more philosophical level. Only about a month before his death in 1955, Einstein wrote: "Death signifies nothing . . . the distinction between past, present, and future is only a stubbornly persistent illusion." As this quotation suggests, relativity raises all sorts of interesting questions about what the passage of time really means. Because these are philosophical questions, they do not have definitive answers, and you will have to decide for yourself what these questions mean to you. But I believe that one thing is clear: Based on our understanding of spacetime, there seems no getting around the idea that events in spacetime have a permanence to them that cannot be taken away. Once an event occurs, it in essence becomes part of the fabric of our universe. Your life is a series of events, and this means that when you put them all together, you are creating your own, indelible mark on the universe. Perhaps if everyone understood that, we might all be a little more careful to make sure that the mark we leave is one that we are proud of.

ACKNOWLEDGMENTS

I COULD not have written this book without the contributions of many other people. I'd particularly like to thank Mark Voit and Megan Donahue, my coauthors (along with Nick Schneider) on an astronomy textbook, *The Cosmic Perspective*. Mark and Megan—both of whom actually know much more about relativity than I do—helped me write the chapters on relativity in our textbook, and many of the examples and analogies that I've presented in this book are ones that we originally created for the textbook. In addition, I received tremendous help both from Mark Voit and from University of Colorado professor Andrew Hamilton, each of whom read this entire manuscript carefully and gave me suggestions on how to keep the level appropriate to a general audience while still retaining scientific accuracy. I also received a great many excellent suggestions for improving clarity from two careful nonscientist readers: my good friend Joan Marsh and my high school son, Grant Bennett.

Many others among my teachers and colleagues have also been very important in shaping my understanding of relativity. I'd particularly like to acknowledge T. M. Helliwell, from whom I took my first course in relativity at Harvey Mudd College, and five professors at the University of Colorado: Andrew Hamilton, J. Michael Shull, Richard McCray, Theodore P. Snow, and J. McKim Malville. I'd also like to acknowledge several books that have been especially important both in helping me understand the content of relativity and in giving me ideas about how to teach it to the public; indeed, many of the thought experiments and analogies I've offered in this book have their roots in examples I first encountered in these other books, which include: Einstein's own book for the general public, simply titled *Relativity*; *The Relativity Explosion*, by Martin Gardner; *Spacetime Physics*, by Edwin Taylor and John Archibald Wheeler; *The Physical Universe*, by Frank Shu; *Gravity and Spacetime*, by John Archibald Wheeler; *Gravity's Fatal Attraction*, by Mitchell Begelman and Martin Rees; *Black Holes and Time Warps*, by Kip S. Thorne; *Cosmos*, by Carl Sagan; and *Einstein: His Life and Universe*, by Walter Isaacson.

Special thanks to everyone at Columbia University Press, especially the editorial team of Patrick Fitzgerald and Bridget Flannery-McCoy, for putting their faith into this project and turning my manuscript into a published book. I also thank Nancy Whilton and others at Pearson Addison-Wesley for allowing me to write this book with a significant amount of material—including most of the illustrations—adapted from our textbook *The Cosmic Perspective*. Finally, I thank my wife, Lisa, and my children, Grant and Brooke, for their ongoing support, inspiration, and insights.

INDEX